for

Algebra

Boca Raton, Florida

ISBN 13: 9781423202547

ISBN 10: 1423202546

Author: Dr. S. B. Kizlik

Publisher:

BarCharts, Inc.

6000 Park of Commerce Boulevard, Suite D

Boca Raton, FL 33487

www.quickstudy.com

Printed in Thailand

Contents

Study Hints

NOTE TO STUDENT:

Use this QuickStudy® booklet to make the most of your studying time.

All equations are set in boldface type for easy reference.

EX: $\sqrt{x^2+4} \neq x+2$, BUT $\sqrt{x^2+4x+4} = \sqrt{(x+2)^2}$

QuickStudy® notes provide need-to-know information; read them carefully to better understand key concepts.

NOTES
An understanding of the **Operations** section of **Rational Expressions** is required to work "complex fractions."

Take your learning to the next level with QuickStudy®!

1 Set Theory

Notation

- { } braces indicate the beginning and end of a set notation; when listed, elements or members must be separated by commas; sets are finite (ending, or having a last element) unless otherwise indicated.
  EX: A = {4, 8, 16}

- ... indicates continuation of a pattern.
  EX: B = {5, 10, 15, ... , 85, 90}

- ... at the end indicates an **infinite set**, that is, a set with no last element.
  EX: C = {3, 6, 9, 12, ...}

- | is a symbol which literally means "such that."

- $\in$ means "is a member of" OR "is an element of."
  EX: If **A = {4, 8, 12}** then **12** $\in$ **A** because **12** is in set **A**.

- $\notin$ means "is not a member of" OR "is not an element of."
  EX: If **B = {2, 4, 6, 8}** then **3** $\notin$ **B** because **3** is not in set **B**.

- $\varnothing$ means **empty set** OR **null set**; a set containing no elements or members, but which is a subset of all sets; also written as **{ }**.

- $\subset$ means "is a subset of"; also written as $\subseteq$.

- $\not\subset$ means "is not a subset of"; also written as $\not\subseteq$.

- **A**$\subset$**B** indicates that every element of set **A** is also an element of set **B**.

  EX: If A = {3, 6} and **B = {1, 3, 5, 6, 7, 9}** then **A**$\subset$ **B** because the **3** and **6** which are in set **A** are also in set **B**.

- 2^n is the number of subsets of a set when **n** equals the number of elements in that set.
  EX: If **A** = **{4, 5, 6}** then set **A** has **8** subsets because **A** has **3** elements and $2^3 = 8$.

Operations

- $\cup$ means **union**.

- **A**$\cup$**B** indicates the union of set **A** with set **B**; every element of this set is either an element of set **A** OR an element of set **B**; that is, to form the union of two sets, put all of the elements of both sets together into one set making sure not to write any element more than once.

  EX: If **A = {2,4} and B = {4, 8, 16}** then **A**$\cup$**B** = **{2, 4, 8, 16}**.

- $\cap$ means **intersection**.

- $\mathbf{A} \cap \mathbf{B}$ indicates the intersection of set **A** with set **B**; every element of this set is also an element of **BOTH** set **A** and set **B**; that is, to form the intersection of two sets, list only those elements which are found in **BOTH** of the two sets.

  EX: If $\mathbf{A} = \{\mathbf{2,4}\}$ and $\mathbf{B} = \{\mathbf{4, 8, 16}\}$ then $\mathbf{A} \cap \mathbf{B} = \{\mathbf{4}\}$.

- $\overline{\mathrm{A}}$ indicates the **complement** of set **A;** that is, all elements in the universal set which are **NOT** in set **A.**
  EX: If the Universal set is the set of Integers and $\mathbf{A} = \{\mathbf{0, 1, 2, 3, ...}\}$ then $\overline{\mathrm{A}} = \{\mathbf{-1, -2, -3, -4, ...}\}$.

  $\mathrm{A} \cap \overline{\mathrm{A}} = \varnothing$.

Properties

- **A** = **B** means all of the elements in set **A** are also in set **B** and all elements in set **B** are also in set **A**, although they do not have to be in the same order.
  EX: If **A** = **{5, 10}** and **B** = **{10, 5}** then **A** = **B**.

- $\boldsymbol{n}(\mathrm{A})$ indicates the number of elements in set **A**.
  EX: If **A** = **{2, 4, 6}** then $\boldsymbol{n}(\mathbf{A}) = \mathbf{3}$.

- ~ means "**is equivalent to**"; that is, set **A** and set **B** have the same number of elements although the elements themselves may or may not be the same.
  EX: If **A** = **{2, 4, 6}** and **B** = **{6, 12, 18}** then **A ~ B** because $\boldsymbol{n}(\mathbf{A}) = \mathbf{3}$ and $\boldsymbol{n}(\mathbf{B}) = \mathbf{3}$.

- $A \cap B = \varnothing$ indicates **disjoint sets** which have no elements in common.

Sets of Numbers

- **Natural or Counting numbers = {1, 2, 3, 4, 5, ..., 11, 12, ...}**

- **Whole numbers = {0, 1, 2, 3,...., 10, 11, 12, 13, ...}**

- **Integers = {... , -4, -3, -2, -1, 0, 1, 2, 3, 4, ...}**

- **Rational numbers** = {**p/q** | **p** and **q** are integers, **q** $\neq$ **0**}; the sets of Natural numbers, Whole numbers, and Integers, as well as numbers which can be written as proper or improper fractions, are all subsets of the set of Rational numbers.

- **Irrational numbers** = { **x** | **x** is a Real number but is not a Rational number}; the sets of Rational numbers and Irrational numbers have no elements in common and are therefore disjoint sets.

- **Real numbers** = {**x** | **x** is the coordinate of a point on a number line}; the union of the set of Rational numbers with the set of Irrational numbers equals the set of Real numbers.

- **Imaginary numbers** = {***ai*** | **a** is a Real number and ***i*** is the number whose square is **-1**}; $i^2 = -1$ the sets of Real numbers and Imaginary numbers have no elements in common and are therefore disjoint sets.

- **Complex numbers** = {**a** + **b***i* | **a** and **b** are Real numbers and ***i*** is the number whose square is **-1**}; the set of Real numbers and the set of Imaginary numbers are both subsets of the set of Complex numbers.
  EX: 4 + 7*i* ; 3 - 2*i*

2 Properties of Real Numbers

For Any Real Numbers a, b & c

Addition Properties

Closure $a + b$ is a Real number

Commutative $a + b = b + a$

Associative $(a + b) + c = a + (b + c)$

Identity $0 + a = a$ and $a + 0 = a$

Inverse $a + (-a) = 0$ and $(-a) + a = 0$

Multiplication Properties

Closure ab is a Real number

Commutative $ab = ba$

Associative $(ab)c = a(bc)$

Identity $a \bullet 1 = a$ and $1 \bullet a = a$

Inverse $a \bullet {}^{1}/_{a} = 1$ and ${}^{1}/_{a} \bullet a = 1$ if $a \neq 0$

Distributive Property

$a(b + c) = ab + ac$; $a(b - c) = ab - ac$

Properties of Equality

For Any Real Numbers a, b & c

Reflexive:
$a = a$

Symmetric:
If $a = b$ then $b = a$

Transitive:
If $a = b$ and $b = c$ then $a = c$

Addition Property:
If $a = b$ then $a + c = b + c$

Multiplication Property:
If $a = b$ then $ac = bc$

Multiplication Property of Zero:
$a \cdot 0 = 0$ and $0 \cdot a = 0$

Double Negative Property:
$-(-a) = a$

Properties of Inequality

For Any Real Numbers a, b & c

Trichotomy: Either $a > b$, or $a = b$, or $a < b$

Transitive: If $a < b$, and $b < c$ then $a < c$
If $a > b$, and $b > c$ then $a > c$

Addition Property of Inequalities:
If $a < b$ then $a + c < b + c$
If $a > b$ then $a + c > b + c$

Multiplication Property of Inequalities:

If $\mathbf{c} \neq \mathbf{0}$ and $\mathbf{c} > \mathbf{0}$, and $\mathbf{a} > \mathbf{b}$ then $\mathbf{ac} > \mathbf{bc}$;
also, if $\mathbf{a} < \mathbf{b}$ then $\mathbf{ac} < \mathbf{bc}$

If $\mathbf{c} \neq \mathbf{0}$ and $\mathbf{c} < \mathbf{0}$, and $\mathbf{a} > \mathbf{b}$ then $\mathbf{ac} < \mathbf{bc}$;
also, if $\mathbf{a} < \mathbf{b}$ then $\mathbf{ac} > \mathbf{bc}$

3 Operations of Real Numbers

Absolute Value

- $|x| = x$ if x is zero or a positive number; $|x| = -x$ if x is a negative number; that is, the distance (which is always positive) of a number from zero on the number line is the absolute value of that number.
  EX: $|-4| = -(-4) = 4$; $|29| = 29$; $|0| = 0$;
  $|-43| = -(-43) = 43$

Addition

- If the **signs** of the numbers are the **same**: **add** the absolute values of the numbers; the sign of the answer is the same as the signs of the original two numbers.
  EX: $-11 + -5 = -16$ and $16 + 10 = 26$

- If the **signs** of the numbers are **different**: **subtract** the absolute values of the numbers; the answer has the same sign as the number with the larger absolute value.
  EX: $-16 + 4 = -12$ and $-3 + 10 = 7$

Subtraction

- $a - b = a + (-b)$; **subtraction is changed to addition of the opposite number:** that is, change the sign of the second number and follow the rules of addition (never change the sign of the first number since it is the number in back of the subtraction sign that is being subtracted; **14 - 6 ≠ - 14 + - 6)**.

EX: 15 - 42 = 15 + (-42) = -27; - 24 - 5 = - 24 + (-5) = - 29; - 13 - (-45) = -13 + (+45) = 32; - 62 - (-20) = - 62 + (+20) = - 42

Multiplication

- The product of two numbers that have the **same** signs is **positive**.

  EX: (55)(3) = 165;
  (- 30)(- 4) = 120;
  (- 5)(- 12) = 60

- The product of two numbers that have **different** signs is **negative** no matter which number is larger.

  EX: (- 3)(70) = - 210;
  (21)(- 40) = - 840;
  (50)(-3) = - 150

Division

Divisors Do Not Equal Zero

- The quotient of two numbers that have the **same** sign is **positive**.

  EX: (- 14) / (-7) = 2; (44) / (11) = 4

- The quotient of two numbers that have **different** signs is **negative** no matter which number is larger.

  EX: (-24) / (6) = -4; (40) / (-8) = - 5;
  (-14) / (56) = - .25

Double Negative

- **$-(-a) = a$;** that is, the negative sign changes the sign of the contents of the parentheses.
  EX: $-(-4) = 4$; $-(-17) = 17$

4 Algebraic Terms

Combining Like Terms

Adding or Subtracting

- ***$a + a = 2a$; when adding or subtracting terms, they must have exactly the same variables and exponents, although not necessarily in the same order; these are called like terms. The coefficients (numbers in the front) may or may not be the same.***

- **RULE:** combine (add or subtract) only the coefficients of like terms and never change the exponents during addition or subtraction.

  EX: $4xy^3$ and $-7y^3x$ are like terms and may be combined: $4xy^3 + -7y^3x = -3xy^3$.

  Note: only the coefficients were combined and no exponent changed.

  $-15a^2bc$ and $3bca^4$ are not like terms because the exponents of the **a** are not the same in both terms, so they may not be combined.

Multiplying

Product Rule for Exponents

- ***$(a^m)(a^n) = a^{m+n}$; any terms may be multiplied, not just like terms. The coefficients <u>and</u> the variables are multiplied which means the exponents also change.***

- **RULE: multiply the coefficients and multiply the variables** (this means you have to add the exponents of the same variable).
  EX: $(4a^2c)(-12a^3b^2c) = -48a^5b^2c^2$.
  Note: 4 times -12 became -48, a^2 times a^3 became a^5, c times c became c^2, and the b^2 was written down.

Distributive Property for Polynomials

- **Type 1: $a(c + d) = ac + ad$.**
  EX: $4x^3(2xy + y^2) = 8x^4y + 4x^3y^2$
- **Type 2: $(a + b)(c + d) = a(c + d) + b(c + d) = ac + ad + bc + bd$**
  EX: $(2x + y)(3x - 5y) = 2x(3x - 5y) + y(3x - 5y)$
  $= 6x^2 - 10xy + 3xy - 5y^2 = 6x^2 - 7xy - 5y^2$
- **Type 3: $(a + b)(c^2 + cd + d^2) = a(c^2 + cd + d^2) + b(c^2 + cd + d^2) = ac^2 + acd + ad^2 + bc^2 + bcd + bd^2$**
  EX: $(5x + 3y)(x^2 - 6xy + 4y^2) = 5x(x^2- 6xy+4y^2) + 3y(x^2 - 6xy + 4y^2) = 5x^3 - 30x^2y + 20xy^2 + 3x^2y - 18xy^2 + 12y^3 = 5x^3 - 27x^2y + 2xy^2 + 12y^3$

"FOIL" Method for Products of Binomials

- **This is a popular method for multiplying 2 terms by 2 terms only. *FOIL*** means *first times first, outer times outer, inner times inner, and last times last.*
  EX: $(2x + 3y)(x + 5y)$ would be multiplied by multiplying first term times first term, **$2x$** times **$x = 2x^2$;** outer term times outer term, **$2x$** times **$5y = 10xy$;** inner term times inner term, **$3y$** times **$x = 3xy$;** and last term times last term, **$3y$** times **$5y = 15y^2$;** then combining the like terms of **$10xy$** and **$3xy$** gives **$13xy$** with the final answer equaling **$2x^2 + 13xy + 15y^2$.**

Special Products

- **Type 1: $(a + b)^2 = (a + b)(a + b) = a^2 + 2ab + b^2$**
- **Type 2: $(a - b)^2 = (a - b)(a - b) = a^2 - 2ab + b^2$**
- **Type 3: $(a + b)(a - b) = a^2 + 0ab - b^2 = a^2 - b^2$**

Exponent Rules

- **Rule 1: $(a^m)^n = a^{m \cdot n}$; $(a^m)^n$** means the parentheses contents are multiplied **n** times and when you multiply you add exponents.
  EX: $(-2m^4n^2) = (-2m^4n^2)\ (-2m^4n^2)\ (-2m^4n^2) = -8m^{12}n^6$
  Note: The parentheses were multiplied 3 times and then the rules of regular multiplication of terms were used.

- **SHORTCUT RULE:** when raising a term to a power, just multiply exponents.
  EX: $(-2m^4n^2)^3 = -2^3m^{12}n^6$
  Note: The exponents of the **-2, m^4** and **n^2** were all multiplied by the exponent **3**, and that the answer was the same as the example above.

- **CAUTION: $-a^m \neq (-a)^m$**; these two expressions are different.
  EX: $-4yz^2 \neq (-4yz)^2$ because $(-4yz)^2 = (-4yz)(-4yz) = 16y^2z^2$ **while** $-4yz^2$ means $-4 \cdot y \cdot z^2$ and the exponent **2** applies only to the **z** in this situation.

- **Rule 2: $(ab)^m = a^m\ b^m$**
  EX: $(6x^3\ y)^2 = 6^2\ x^6\ y^2 = 36x^6\ y^2$ BUT $(6x^3 + y)^2 = (6x^3 + y)\ (6x^3 + y) = 36x^6 + 12x^3y + y^2$
  Because there is more than one term in the parentheses the distributive property for polynomials must be used in this situation.

◆ **Rule 3:** $\left(\frac{a}{b}\right)^m = \frac{a^m}{b^m}$ when $b \neq 0$;

EX: $\left(\frac{-4x^2y}{5z}\right)^2 = \frac{16x^4y^2}{25z^2}$

◆ **Rule 4:** Zero Power $a^0 = 1$ when $a \neq 0$

Dividing

■ **Quotient Rule:** $\frac{a^m}{a^n} = a^{m-n}$; any terms may be divided, not just like terms; the coefficients and the variables are divided which means the exponents also change.

◆ **RULE:** Divide coefficients and divide variables (this means you have to subtract the exponents of matching variables).

EX: $(-20x^5y^2z) / (5x^2z) = -4x^3y^2$

Note: -20 divided by **5** became **-4,** x^5 divided by x^2 became x^3, and **z** divided by **z** became **1** and therefore did not have to be written because **1** times $-4x^3y^2$ equals $-4x^3y^2$.

■ **Negative Exponent:** $a^{-n} = \frac{1}{a^n}$ **when** $a \neq 0$;

EX: $2^{-1} = \frac{1}{2}$; $(4z^{-3}y^2) / (-3ab^{-1}) = (4y^2b^1) / (-3az^3)$

Note: The **4** and the **-3** both stayed where they were because they both had an invisible exponent of positive **1**; the **y** remained in the numerator and the **a** remained in the denominator because their exponents were both positive numbers; the **z** moved down and the **b** moved up because their exponents were both negative numbers.

5 Steps for Solving a First-Degree Equation with One Variable

- **First, eliminate any fractions** by using the *Multiplication Property of Equality.*

  EX: $\mathbf{\frac{1}{2}(3a + 5) = \frac{2}{3}(7a - 5) + 9}$ would be multiplied on both sides of the = sign by the lowest common denominator of $\frac{1}{2}$ and $\frac{2}{3}$, which is **6**; the result would be $\mathbf{3(3a + 5) = 4(7a - 5) + 54}$

  Note: Only the $\frac{1}{2}$, the $\frac{2}{3}$, and the **9** were multiplied by 6 and not the contents of the parentheses; the parentheses will be handled in the next step which is distribution.

- **Second, simplify the left side of the equation** as much as possible by using the *Order of Operations,* the *Distributive Property*, and *Combining Like Terms*. **Do the same to the right side of the equation.**

  EX: Use distribution first, $\mathbf{3(3a + 5) = 4(7a - 5) + 54}$ would become $\mathbf{9a + 15 = 28a - 20 + 54}$ and then combine like terms to get $\mathbf{9a + 15 = 28a + 34}$.

- **Third, apply the** *Addition Property of Equality* to simplify and organize all terms containing the variable on one side of the equation and all terms which do not contain the variable on the other side.

  EX: $\mathbf{9a + 15 = 28a + 34}$ would become $\mathbf{9a - 28a + 15 - 15 = 28a - 28a + 34 - 15}$ and then combining like terms, $\mathbf{-19a = 19}$.

- **Fourth, apply the** *Multiplication Property of Equality* to make the coefficient of the variable **1**.
  EX: -19a = 19 would be multiplied on both sides by $-\frac{1}{19}$ (or divided by **-19**) to get a **1** in front of the **a** so the equation would become **1a = -1** or simply **a = -1**.

- **Fifth, check the answer** by substituting it for the variable in the original equation to see if it works.

NOTES

- Some equations have exactly one solution (answer). They are **conditional equations**.
  EX: 2k = 18
- Some equations work for all real numbers. They are **identities**.
  EX: 2k = 2k
- Some equations have no solutions. They are **inconsistent equations**.
  EX: 2k + 3 = 2k + 7

6 Steps for Solving a First-Degree Inequality with One Variable

Addition Property of Inequalities

- **For all real numbers a, b, and c, the inequalities $a < b$ and $a + c < b + c$ are equivalent;** that is, any terms may be added to both sides of an inequality and the inequality remains a true statement. This **also applies to $a > b$ and $a + c > b + c$.**

Multiplication Property of Inequalities

- For all real numbers **a, b,** and **c,** with $c \neq 0$ and $c > 0$, the inequalities $a > b$ and $ac > bc$ are equivalent and the inequalities $a < b$ and $ac < bc$ are equivalent; that is, when **c** is a positive number the inequality symbols stay the same as they were before the multiplication.
  EX: If $8 > 3$ then multiplying by **2** would make $16 > 6$, which is a true statement.

- For all real numbers **a, b,** and **c,** with $c \neq 0$ and $c < 0$, the inequalities $a > b$ and $ac < bc$ are equivalent and the inequalities $a < b$ and $ac > bc$ are equivalent; that is, when **c** is a **negative number** the inequality symbols must be reversed from the way they were before the multiplication for the inequality to remain a true statement.
  EX: If $8 > 3$ then multiplying by **-2** would make $-16 > -6$, which is false unless the inequality symbol is reversed to make it true, $-16 < -6$.

Steps for Solving

- **First, simplify the left side of the inequality** in the same manner as an equation, applying the order of operations, the distributive property, and combining like terms. **Simplify the right side in the same manner.**

- **Second, apply the** *Addition Property of Inequality* to get all terms which have the variable on one side of the inequality symbol and all terms which do not have the variable on the other side of the symbol.

- **Third, apply the** *Multiplication Property of Inequality* to get the coefficient of the variable to be a 1; (remember to reverse the inequality symbol when multiplying or dividing by a negative number, this is NOT done when multiplying or dividing by a positive number).

- **Fourth, check the solution** by substituting some numerical values of the variable in the original inequality.

7 Order of Operations

- **First, simplify any enclosure symbols:** parentheses ( ), brackets [ ], braces { } if present:
  - Work the enclosure symbols from the innermost and work outward.
  - Work separately above and below any fraction bars since the entire top of a fraction bar is treated as though it has its own invisible enclosure symbols around it and the entire bottom is treated the same way.
- **Second, simplify any exponents and roots,** working from left to right. **Note:** The $\sqrt{\ }$ symbol is used only to indicate the positive root, except that $\sqrt{0}=0$.
- **Third, do any multiplication and division** in the order in which they occur, working from left to right. **Note:** If division comes before multiplication then it is done first, if multiplication comes first then it is done first.
- **Fourth, do any addition and subtraction** in the order in which they occur, working from left to right. **Note:** If subtraction comes before addition in the problem then it is done first, if addition comes first then it is done first.

8 Factoring

NOTES

Some algebraic polynomials cannot be factored. The following are methods of handling those which can be factored. When the factoring process is complete the answer can always be checked by multiplying the factors out to see if the original problem is the result. That will happen if the factorization is a correct one.

A polynomial is factored when it is written as a product of polynomials with integer coefficients and all of the factors are prime. The order of the factors does not matter.

First Step: "GCF"

- **Factor out the Greatest Common Factor (GCF),** if there is one. The GCF is the largest number which will divide evenly into every coefficient together with the lowest exponent of each variable common to all terms.

  EX: $\mathbf{15a^3c^3 + 25a^2c^4d^2 - 10a^2c^3d}$ has a greatest common factor of $\mathbf{5a^2c^3}$ because **5** divides evenly into **15, 25,** and **10**; the lowest degree of **a** in all three terms is **2**; the lowest degree of **c** is **3**; the GCF is $\mathbf{5a^2c^3}$; the factorization is $\mathbf{5a^2c^3\ (3a + 5cd^2 - 2d)}$.

Second Step: Categorize & Factor

- **Identify the problem as belonging in one of the following categories.** Be sure to place the terms in the correct order first: highest degree term to the lowest degree term.
  EX: $-2A^3 + A^4 + 1 = A^4 - 2A^3 + 1$

CATEGORY	FORM OF PROBLEM	FORM OF FACTORS
TRINOMIALS (3 TERMS)	$ax^2 + bx + c$ *($a \neq 0$)*	If $a = 1$: $(x + h)(x + k)$ where $h \cdot k = c$ and $h + k = b$; h and k may be either positive or negative numbers. If $a \neq 1$: $(mx + h)(nx + k)$ where $m \cdot n = a$, $h \cdot k = c$, and $h \cdot n + m \cdot k = b$; m, h, n and k may be either positive or negative numbers. Trial and error methods may be needed. (see *Special Factoring Hints*)
	$x^2 + 2cx + c^2$ *(perfect square)*	$(x + c)(x + c) = (x + c)^2$ where c may be either a positive or a negative number
BINOMIALS (2 TERMS)	$a^2x^2 - b^2y^2$ *(difference of 2 squares)*	$(ax + by)(ax - by)$
	$a^2x^2 + b^2y^2$ *(sum of 2 squares)*	PRIME -- cannot be factored!
	$a^3x^3 + b^3y^3$ *(sum of 2 cubes)*	$(ax + by)(a^2x^2 - abxy + b^2y^2)$
	$a^3x^3 - b^3y^3$ *(difference of 2 cubes)*	$(ax - by)(a^2x^2 + abxy + b^2y^2)$ (see *Special Factoring Hints*)
PERFECT CUBES (4 TERMS)	$a^3x^3 + 3a^2bx^2 + 3ab^2x + b^3$	$(ax + b)^3 = (ax + b)(ax + b)(ax + b)$
	$a^3x^3 - 3a^2bx^2 + 3ab^2x - b^3$	$(ax - b)^3 = (ax - b)(ax - b)(ax - b)$
	$ax + ay + bx + by$ *(2 - 2 grouping)*	$a(x + y) + b(x + y) = (x + y)(a + b)$
GROUPING	$x^2 + 2cx + c^2 - y^2$ *(3 - 1 grouping)*	$(x + c)^2 - y^2 = (x + c + y)(x + c - y)$
	$y^2 - x^2 - 2cx - c^2$ *(1 - 3 grouping)*	$y^2 - (x + c)^2 = (y + x + c)(y - x - c)$

Special Factoring Hints

Trinomials

◆ Where the coefficient of the highest degree term is not 1

- The first term in each set of parentheses must multiply to equal the first term (highest degree) of the problem. The second term in each set of parentheses must multiply to equal the last term in the problem. The middle term must be checked on a trial and error basis using: outer times outer plus inner times inner; $ax^2 + bx + c = (mx + h)(nx + k)$ where mx times nx equals ax^2, h times k equals c, and mx times k plus h times nx equals bx.

  EX: To factor $3x^2 + 17x - 6$ all of the following are possible correct factorizations:
  $(3x + 3)(x - 2)$; $(3x + 2)(x - 3)$;
  $(3x + 6)(x - 1)$; $(3x + 1)(x - 6)$.
  Note: The only set which results in a **17x** for the middle term when applying "outer times outer plus inner times inner" is the last one, $(3x + 1)(x - 6)$. It results in $-17x$ and $+17x$ is needed, so both signs must be changed to get the correct middle term. Therefore, the correct factorization is $(3x - 1)(x + 6)$.

Binomials

◆ The sum or difference of two cubes

- This type of problem, $a^3x^3 \pm b^3y^3$, requires the memorization of the following procedure: The factors are two sets of parentheses with **2** terms in the first set and **3** terms in the second. To find the **2** terms in the first set of parentheses take the cube root of both of the terms in the problem and join them by the same middle sign found in the problem. The **3** terms in the second set of parentheses are generated from the **2** terms in the first set of parentheses. The first term in the second set of parentheses is the square of the first term in the first set of parentheses; the last term in the second set is the square of the last term in the first set; the middle term of the second set of parentheses is found by multiplying the first term and the second term from the first set of parentheses together and changing the sign. Thus, $\mathbf{a^3x^3 \pm b^3y^3 = (ax \pm by)(a^2x^2 \mp abxy + b^2y^2)}$.

EX: To factor $\mathbf{27x^3 - 8}$ find the first set of parentheses to be $\mathbf{(3x - 2)}$ because the cube root of $\mathbf{27x^3}$ is $\mathbf{3x}$ and the cube root of **8** is **2**. Find the **3** terms in the second set of parentheses by squaring $\mathbf{3x}$ to get $\mathbf{9x^2}$; square the last term **-2** to get **+4**; and to find the middle term multiply $\mathbf{3x}$ times **-2** and change the sign to get $\mathbf{+6x}$. Therefore, the final factorization of $\mathbf{27x^3 - 8}$ is $\mathbf{(3x - 2)(9x^2 + 6x + 4)}$.

Perfect Cubes (4 Terms)

- Perfect cubes, such as $\mathbf{a^3x^3 \pm 3a^2bx^2 + 3ab^2x \pm b^3}$, factor into three sets of parentheses, each containing exactly the same two terms; therefore, the final factorization is written as one set of parentheses to the third power, thus a perfect cube: $\mathbf{(ax \pm b)^3 = a^3x^3 \pm 3a^2bx^2 + 3ab^2x \pm b^3}$.

EX: To factor $\mathbf{27x^3 - 54x^2 + 36x - 8}$ it must be first observed that the problem is in correct order and that it is a perfect cube; then the answer is simply the cube roots of the first term and the last term placed in a set of parentheses to thc third power; so the answer to this example is $\mathbf{(3x - 2)^3}$.

9 Rational Expressions

NOTES

The quotient of two polynomials where the denominator cannot equal zero is a rational expression.

EX: $\frac{(x+4)}{(x-3)}$ where $x \neq 3$, since **3** would make the denominator, **x - 3**, equal to zero.

Basics

- **Domain:** set of all Real numbers which can be used to replace a variable.

  EX: The domain for the rational expression $\frac{(x+5)(x-2)}{(x+1)(4-x)}$ is $\{x|x \in \text{Reals and } x \neq -1 \text{ or } x \neq 4\}$.

  - That is, **x** can be any Real number except **-1** or **4** because **-1** makes **(x + 1)** equal to zero and **4** makes **(4 - x)** equal to zero; therefore, the denominator would equal zero, **which it must not**.
  - Notice that numbers which make numerator equal to zero, **-5** and **2**, are members of the domain since fractions may have zero in numerator but not in denominator.

- **Rule 1:**
  - If **x/y** is a rational expression then **x/y = xa/ya** when $a \neq 0$.

- That is, you may multiply a rational expression (or fraction) by any non-zero value as long as you multiply both numerator and denominator by the same value.
  > Equivalent to multiplying by **1** since **a/a=1**.
  EX: (x/y)(1) = (x/y)(a/a) = xa/ya
  Note: 1 is equal to any fraction which has the same numerator and denominator.

Rule 2:

- If $\frac{xa}{ya}$ is a rational expression, $\frac{xa}{ya}=\frac{x}{y}$ when $a \neq 0$.
  - That is, you may write a rational expression in lowest term because $\frac{xa}{ya}=\left(\frac{x}{y}\right)\left(\frac{a}{a}\right)=\left(\frac{x}{y}\right)(1)=\frac{x}{y}$ since $\frac{a}{a}=1$

Lowest Terms:

- Rational expressions are in lowest terms when they have no common factors other than **1.**
- **Step 1:** Completely **factor** both numerator and denominator.
- **Step 2: Divide** both the numerator and the denominator by the greatest common factor or by the common factors until no common factors remain.

EX: $\frac{(x^2+8x+15)}{(x^2+3x-10)} = \frac{(x+5)(x+3)}{(x+5)(x-2)} = \frac{(x+3)}{(x-2)}$

because the common factor of **(x + 5)** was divided into the numerator and the denominator since $\frac{(x+5)}{(x+5)}=1$.

NOTES

Only factors can be divided into both numerators and denominators, **never terms**.

Operations

Addition (denominators must be the same)

◆ Rule 1:

- If **a/b** and **c/b** are rational expressions and **b ≠ 0**, then: $\frac{a}{b}+\frac{c}{b}=\frac{(a+c)}{b}$.
  > If denominators are already the same, simply add numerators and write this sum over common denominator.

◆ Rule 2:

- If **a/b** and **c/d** are rational expressions and **b ≠ 0** and **d ≠ 0**, then: $\frac{a}{b}+\frac{c}{d}=\frac{(ad)}{(bd)}+\frac{(cb)}{(bd)}=\frac{(ad+cb)}{(bd)}$
  > If denominators are not the same they must be made the same before numerators can be added.

◆ Addition Steps

- If the **denominators are the same**, then:
  > **Add the numerators.**
  > **Write answer** over common denominator.
  > Write **final answer** in lowest terms, making sure to follow directions for finding lowest terms as indicated above.

EX: $\frac{(x+2)}{(x-6)}+\frac{(x-1)}{(x-6)}=\frac{(2x+1)}{(x-6)}$

- If the **denominators are not the same,** then:
  > Find the **least common denominator.**
  > **Change** all of the rational expressions so they have the same common denominator.
  > **Add numerators.**
  > **Write the sum** over the **common denominator.**
  > Write the **final answer** in lowest terms.

$$\frac{x+3}{x+5}+\frac{x+1}{x-1}=\frac{(x+3)(x-1)}{(x+5)(x-1)}+\frac{(x+1)(x+5)}{(x-1)(x+5)}=$$

$$\frac{2x^2+8x+2}{x^2+4x-5}$$

Note: If denominators are of a degree greater than one, try to factor all denominators first so the least common denominator will be the product of all different factors from each denominator.

Subtraction (denominators must be the same)

◆ Rule 1:

- If **a/b** and **c/b** are rational expressions and **b ≠ 0**, then: $\left(\frac{a}{b}\right)-\left(\frac{c}{b}\right)=\left(\frac{a}{b}\right)+\left(\frac{-c}{b}\right)=\frac{(a-c)}{b}$
  > **If denominators are the same** be sure to change all signs of the terms in numerator of rational expression which is behind (to the right of) subtraction sign; then add numerators and write result over common denominator.

EX: $\frac{x-3}{x+1}-\frac{x+7}{x+1}=\frac{x-3+(-x)+(-7)}{x+1}=\frac{-10}{x+1}$

◆ **Rule 2:**

- If **a/b** and **c/d** are rational expressions and $b \neq 0$ and $d \neq 0$, then:

$$\frac{a}{b}-\frac{c}{d}=\frac{(ad)}{(bd)}-\frac{(cb)}{(bd)}=\frac{(ad-cb)}{bd}$$

> **If denominators are not the same** they must be made the same before numerators can be subtracted. Be sure to change signs of all terms in numerator of rational expression which follows subtraction sign after rational expressions have been made to have a common denominator. Combine numerator terms and write result over common denominator.
>
> **Note:** When denominators of rational expressions are additive inverses (opposite signs), then signs of all terms in denominator of expression behind subtraction sign should be changed. This will make denominators the same and terms of numerators can be combined as they are. Subtraction of rational expressions is changed to addition of opposite of either numerator (most of time) or denominator (most useful when denominators are additive inverses) but never both.

◆ **Subtraction Steps:**

- **If denominators are the same,** then:

> **Change signs** of all terms in numerator of a rational expression which follows any subtraction sign.
>
> **Add the numerators.**
>
> **Write answer** to this addition over common denominator.

> Write **final answer** in lowest terms, making sure to follow directions for finding lowest terms as indicated above.

EX: $\frac{(x+2)}{(x-6)} - \frac{(x-1)}{(x-6)} = \frac{[x+2+(-x)+1]}{(x-6)} = \frac{3}{(x-6)}$

- **If denominators are not the same,** then:

> **Find** the least common denominator.

> **Change** all of the rational expressions so they have the same common denominator.

> **Multiply** factors in the numerators if there are any.

> **Change** the signs of all of the terms in the numerator of any rational expressions which are behind subtraction signs.

> **Add numerators.**

> **Write the sum** over the common denominator.

> Write the **final answer** in lowest terms.

EX: $\frac{(x+3)}{(x+5)} - \frac{(x+1)}{(x-1)} = \frac{(x+3)(x-1)}{(x+5)(x-1)} - \frac{(x+1)(x+5)}{(x-1)(x+5)} = \frac{-4x-8}{x^2+4x-5}$

NOTES

If denominators are of a degree greater than one, try to factor all denominators first, so the least common denominator will be the product of all different factors from each denominator.

Multiplication (denominators do not have to be the same)

◆ Rule:

- If **a, b, c,** and **d** are Real numbers and **b** and **d** are non-zero numbers, then: $\left(\frac{a}{b}\right)\left(\frac{c}{d}\right)=\frac{(ac)}{(bd)}$; [top times top and bottom times bottom]

◆ Multiplication Steps:

- Completely **factor** all numerators and denominators.
- **Write** problem as one big fraction with all numerators written as factors (multiplication indicated) on top and all denominators written as factors (multiplication indicated) on bottom.
- **Divide** both numerator and denominator by all of the common factors; that is, write in lowest terms.
- **Multiply** the remaining factors in the numerators together and write the result as the final numerator.
- **Multiply** the remaining factors in the denominators together and write the result as the final denominator.

EX: $\left(\frac{x+3}{x^2+2x+1}\right)\left(\frac{x^2-2x-3}{x^2-9}\right)=$

$$\frac{(x+3)}{(x+1)(x+1)}\cdot\frac{(x-3)(x+1)}{(x+3)(x-3)}=\frac{1}{x+1}$$

Division

◆ Definition

- Reciprocal of a rational expression

  $\frac{x}{y}$ is $\frac{y}{x}$ because $\left(\frac{x}{y}\right)\left(\frac{y}{x}\right)=1$,

  [reciprocal may be found by inverting the expression].

  EX: The reciprocal of $\left(\frac{x-3}{x+7}\right)$ is $\left(\frac{x+7}{x-3}\right)$

◆ Rule

- If **a, b, c,** and **d** are Real numbers and **a, b, c,** and **d** are non-zero numbers, then:

  $\frac{a}{b} \div \frac{c}{d} = \left(\frac{a}{b}\right)\left(\frac{d}{c}\right) = \left(\frac{ad}{bc}\right)$

◆ Division Steps

- **Reciprocate** (flip) rational expression found behind division sign (immediately to right of division sign).
- **Multiply** resulting rational expressions, making sure to follow steps for multiplication as listed above.

  EX: $\frac{x^2-2x-15}{x^2-10x+25} \div \frac{(x+2)}{(x-5)} =$

  $\frac{x^2-2x-15}{x^2-10x+25} \cdot \frac{(x-5)}{(x+2)}$

  Numerators and denominators would then be factored, written in lowest terms, and yield a final answer of $\frac{(x+3)}{(x+2)}$.

10 Rational Expressions in Equations

- **Definition:** Equations containing rational expressions are **algebraic fractions**.
- **Steps:**
  - **Determine least common denominator** for all rational expressions in equation.
  - Use the *Multiplication Property of Equality* to **multiply** all terms on both sides of the equality sign by the common denominator and thereby **eliminate** all algebraic fractions.
  - **Solve** resulting equation using appropriate steps, depending on degree of equation which resulted from following the above bullet.
  - **Check** answers because numerical values which cause denominators of rational expressions in original equation to be equal to zero are extraneous solutions, not true solutions of original equation.

11 Complex Fractions

NOTES

An understanding of the **Operations** section of **Rational Expressions** is required to work "complex fractions."

- **Definition:** A rational expression having a fraction in the numerator or denominator or both is a **complex fraction**.

  EX: $\dfrac{x - \frac{1}{x}}{x}$

- **Two Available Methods:**
  - **Simplify the numerator** (combine rational expressions found only on top of the complex fraction) and denominator (combine rational expressions found only on bottom of the complex fraction) then divide numerator by denominator; that is, multiply numerator by reciprocal (flip) of denominator. OR

  - **Multiply the complex fraction** (both in numerator and denominator) by least common denominator of all individual fractions which appear anywhere in the complex fraction. This will eliminate the fractions on top and bottom of the complex fraction and result in one simple rational expression. Follow steps listed for simplifying rational expressions.

12 Synthetic Division

NOTES

A process used to divide a polynomial by a binomial in the form of $\mathbf{x + h}$ where $\mathbf{h}$ is an integer.

Steps:

- Write the polynomial in descending order [from highest to lowest power of variable].
  EX: $3x^3 - 6x + 2$
- **Write all coefficients** of dividend under long division symbol, making sure to write zeros which are coefficients of powers of variable which are not in polynomial.
  EX: Writing coefficients of polynomial in example above, write **3 0 - 6 2** because a zero is needed for the $\mathbf{x^2}$ since this power of **x** does not appear in polynomial and therefore has a coefficient of zero.
- **Write the binomial** in descending order.
  EX: $x - 2$
- **Write additive inverse of constant term of binomial** in front of long division sign as divisor.
  EX: The additive inverse of the **-2** in the binomial **$x - 2$** is simply **+2;** that is, change the sign of this term.
- **Bring up first number in dividend** so it will become the first number in quotient (the answer).

- **Multiply** number just placed in quotient by divisor, **2.**
  - Add result of multiplication to next number in dividend.
  - Result of this addition is next number coefficient in quotient; so write it over next coefficient in dividend.
- **Repeat previous step** until all coefficients in dividend have been used.
  - Last number in the quotient is the numerator of a remainder which is written as a fraction with the binomial as the denominator.

  EX: In $2\overline{)3\ \ 0\ \ -6\ \ 2}$ bring 3 up over 3; multiply 2 by 3, write product 6 under 0; add, write the sum 6 over 0; multiply 2 by 6, write product 12 under -6; add, write sum 6 over -6; multiply 2 by 6, write product 12 under 2; add, the sum 14 is the remainder; therefore;

$$(3x^3 - 6x + 2) \div (x - 2) = 3x^2 + 6x + 6 + \frac{14}{(x-2)}$$

- First exponent in answer (quotient) is one less than highest power of dividend because division was by a variable to first degree.

13 Roots & Radicals

Basics

- **Definition:** The real number **b** is the ***n*th** root of **a** if $b^n = a$.
- **Radical Notation:** If $n \neq 0$ then $a^{\frac{1}{n}} = \sqrt[n]{a}$ and $\sqrt[n]{a} = a^{\frac{1}{n}}$. The symbol $\sqrt{}$ is the radical or root symbol. The **a** is the radicand. The **n** is the index or order.
- **Special Note:** Equation $a^2 = 4$ has two solutions, **2** and **-2**. However, the radical $\sqrt{a}$ represents only the non-negative square root of **a**.
- **Definition of Square Root:** For any Real number **a**, $\sqrt{a^2} = |a|$, that is, the non-negative numerical value of **a** only.

  EX: $\sqrt{4} = +2$ only, by definition of the square root.

Rules

- **For Any Real Numbers, m and n,** with **m/n** in lowest terms and $n \neq 0$, $a^{\frac{m}{n}} = (a^m)^{\frac{1}{n}} = \sqrt[n]{a^m}$; OR $a^{\frac{m}{n}} = (a^{\frac{1}{n}})^m = (\sqrt[n]{a})^m$.
- **For Any Real Numbers, m and n,** with **m/n** in lowest terms and $n \neq 0$, $a^{-\frac{m}{n}} = \frac{1}{a^{\frac{m}{n}}}$.
- **For Any Non-Zero Real Number n,** $(a^n)^{\frac{1}{n}} = a^1 = a$; OR $(a^{\frac{1}{n}})^n = a^1 = a$

- **For Real Numbers a** and **b** and natural number **n,**
  $(\sqrt[n]{a}\sqrt[n]{b})=\sqrt[n]{ab}$; OR $\sqrt[n]{ab}=\sqrt[n]{a}\cdot\sqrt[n]{b}$
  i.e., as long as the radical expressions have the same index **n,** they may be multiplied together and written as one radical expression of a product OR they may be separated and written as the product of two or more radical expressions; the radicands do not have to be the same for multiplication.
- **For Real Numbers a** and **b,** and natural number **n,**
  $\frac{\sqrt[n]{a}}{\sqrt[n]{b}}=\sqrt[n]{\frac{a}{b}}$; OR $\sqrt[n]{\frac{a}{b}}=\frac{\sqrt[n]{a}}{\sqrt[n]{b}}$ i.e., as long as the radical expressions have the **same index** they may be written as one quotient under one radical symbol OR they may be separated and written as one radical expression over another radical expression; the radicands do not have to be the same for division.
- **Terms containing radical expressions cannot be combined** unless they are like or similar terms and the radical expressions which they contain are the same; the **indices and radicands must be the same for addition and subtraction.**

  EX: $3x\sqrt{2}+5x\sqrt{2}=8x\sqrt{2}$

  BUT $3y\sqrt{5}+7y\sqrt{3}$ **cannot** be combined because the radical expressions they contain are not the same.

  The terms $7m\sqrt{2}$ and $8m\sqrt[3]{2}$ **cannot** be combined because the indices (plural of index) are not the same.

Simplifying Radical Expressions

- **When the radical expression contains one term and no fractions** $(\sqrt{12m^2})$ then:
  - Take the greatest root of the coefficient.

    EX: For $\sqrt{32}$ use $\sqrt{16} \cdot \sqrt{2}$, not $\sqrt{4} \cdot \sqrt{8}$, because $\sqrt{8}$ is not in simplest form.
  - Take the greatest root of each variable in the term. Remember $\sqrt[n]{a^n} = a$; that is, the power of the variable is divided by the index.
    - This is accomplished by first noting if the power of the variable in the radicand is less than the index. If it is, the radical expression is in its simplest form.
    - If the power of the variable is not less than the index, divide the power by the index. The quotient is the new power of the variable to be written outside of the radical symbol. The remainder is the new power of the variable still written inside of the radical symbol.

    EX: $\sqrt[3]{a^7} = a^2\sqrt[3]{a}$; $\sqrt[3]{8ab^5} = 2b\sqrt[3]{ab^2}$

- **When the radical expression contains more than one term and no fractions** $(\sqrt{x^2+6x+9})$ then:
  - **Factor**, if possible, and take the **root of the factors.** Never take the root of individual terms of a radicand.

  EX: $\sqrt{x^2+4} \neq x+2$, BUT $\sqrt{x^2+4x+4} =$ $\sqrt{(x+2)^2}$

◆ If the radicand is not factorable, then the radical expression cannot be simplified because you cannot take the root of the terms of a radicand.

When the radical expression contains fractions

◆ If the fraction(s) is part of one radicand, under the radical symbol $\left(\sqrt{\frac{x}{3}}\right)$ then:

- **Simplify the radicand** as much as possible to make the radicand one Rational expression so it can be separated into the root of the numerator over the root of the denominator. Simplify the radical expression in the numerator. Simplify the radical expression in the denominator.
- **Never leave a radical expression in the denominator.** It is not considered completely simplified until the fraction is in lowest terms. **Rationalize** the expression to remove the radical expression from the denominator as follows:
  - > **Step 1: Multiply** the numerator and the denominator by the radical expression needed to eliminate the radical expression from the denominator. A radical expression in the numerator is acceptable.

    EX: $\frac{(5x\sqrt{2})}{(7\sqrt{3})}$ must be multiplied by $\frac{(\sqrt{3})}{(\sqrt{3})}$ so the denominator becomes **21** with no radical symbols in it. The numerator becomes $\mathbf{5x\sqrt{6}}$.
  - > **Step 2:** Write the answer in lowest terms.

◆ If the fraction contains monomial radical expressions, $\left(\frac{(x)}{(\sqrt{5})}\right)$ then:

- **If the radical expression is in the numerator only**, simplify it and write the fraction in lowest terms.
- **If the radical expression is in the denominator only**, rationalize the fraction so no radical symbols remain there. Simplify the resulting fraction to lowest terms.
- **If radical expressions are in both numerator and denominator**, either:
  > Simplify each separately, rationalize the denominator and write the answer in lowest terms, OR
  > Make the indices on all radical symbols the same, put the numerator and the denominator under one common radical symbol, write in lowest terms, separate again into a radical expression in the numerator and a radical expression in the denominator, rationalize the denominator, and write the answer in lowest terms.

◆ If the radical expressions are part of polynomials in a rational expression $\left(\frac{(x+\sqrt{2})}{(3x+\sqrt{5})}\right)$ then:

- **If the radical expressions are not in the denominator,** then simplify the fraction and write the answer in lowest terms.

- **If the radical expressions are in the denominator,** rationalize it by multiplying the numerator and the denominator by the conjugate of the denominator. **Conjugates are expressions with the middle sign changed.**

  EX: The conjugate of $3x+5\sqrt{2}$ is $3x-5\sqrt{2}$ because when they are multiplied, the radical symbol is eliminated.

14 Radical Operations

- **Addition and Subtraction:** Only radical expressions which have the same index and the same radicand may be added.
- **Multiplication and Division:**
  - **Monomials** may be multiplied when the indices are the same even though the radicands are not.

    EX: $(3x\sqrt{5})(2\sqrt{7}) = 6x\sqrt{35}$
  - **Binomials** may be multiplied using any method for multiplying regular binomial expressions if indices are the same. (e.g. FOIL).

    EX: $(9m+2\sqrt{5})(3m-5\sqrt{7}) = 27m^2 - 45m\sqrt{7}$

    $+ 6m\sqrt{5} - 10\sqrt{35}$
  - Other **polynomials** are multiplied using the distributive property for multiplication if the indices are the same.
  - Division may be simplification of radical expressions or multiplication by the reciprocal of the divisor. Rationalize the answer so it is in lowest terms without a radical expression in the denominator.

15 Radical Expressions in Equations

- **Rule: Both sides of an equation may be raised to same power. Caution:** Since both entire sides must be raised to the same power, place each side in a separate set of parentheses first.
- **Steps**:
  - If the equation has only one radical expression **EX:** $\sqrt{3x}+5=x$ then:
    - **Isolate** the radical expression on one side of the equal sign.
    - **Raise** both sides of the equation to the same power as the index.
    - **Solve** the resulting equation.
    - **Check** the solution(s) in the original equation because extraneous solutions are possible.

    EX: $\sqrt{3x}+5=x$ becomes $\sqrt{3x}=x-5$, then squaring both sides gives $\mathbf{3x = x^2 - 10x + 25}$ because when the entire right side, which is a binomial, $\mathbf{x - 5}$, is squared, $\mathbf{(x - 5)^2}$, the result is $\mathbf{x^2 - 10x + 25}$. This is now a second degree equation. The steps for solving a quadratic equation should now be followed.

- With equations containing two radical expressions
  EX: $\sqrt{x} + \sqrt{2x} = 4$
  - **Change** the radical expressions to have the same index.
  - **Separate** the radical expressions, placing one on each side of the equal sign.
  - **Raise** both sides of the equation to the same power as the index.
  - **Repeat** the 2 steps above until all radical expressions are eliminated.
  - **Solve** the resulting equation and check the solution(s).
- If the equation has more than two radicals:
  - **Change** the radical expression to the same index.
  - **Separate** as many radical expressions as possible on different sides of the equal sign.
  - **Raise** both sides of the equation to the power of the index.
  - **Repeat** the 2 steps above until all radical expressions are eliminated.
  - **Solve** the resulting equation and check the solution(s).

16 Quadratic Equations

- **Definition:** Second-degree equations in one variable which can be written in the form $\mathbf{ax^2 + bx + c = 0}$ where **a, b,** and **c** are Real numbers and $\mathbf{a \neq 0}$.
- **Property:** If **a** and **b** are Real numbers and **(a)(b) = 0** then either **a = 0** or **b = 0** or both equal zero. At least one of the factors has to be equal to zero.
- **Steps:**
  - ◆ Set the equation **equal to zero.** Combine like terms. Write in descending order.
  - ◆ **Factor**; (if factoring is not possible then go to the quadratic formula below)
    - Set each factor **equal to zero.** See above: "If a product is equal to zero at least one of the factors must be zero."
    - **Solve** each resulting equation and check the solution(s).
  - ◆ Use the **quadratic formula** if factoring is not possible.
    - The quadratic formula is: $x = \frac{-b \pm \sqrt{b^2 - 4ac}}{2a}$
    - **a, b,** and **c** come from the second-degree equation which is to be solved. After the second-degree equation has been set equal to zero, **a** is the coefficient (number in front of) of the second-degree term, **b** is the coefficient (number in front of) of the first-degree term (if no first-degree term is present then **b** is zero), and **c** is the constant term (no variable showing). **Note:** $\mathbf{ax^2 + bx + c = 0}$.

- **Substitute** the numerical values for **a, b,** and **c** into the quadratic formula.
- **Simplify** completely.
- **Write the two answers**, one with + in front of the radical expression, and one with – in front of the radical expression in the formula. Complete any additional simplification to get the answers in the required form.

17 Complex Numbers

- **Definition:** The set of all numbers, $\mathbf{a + b}i$, where **a** and **b** are Real numbers and $i^2 = -1$; that is, i is the number whose square equals **-1** or $i = \sqrt{-1}$.

  NOTE: $i^2 = -1$ will be used in multiplication and division. The complex number $3 + 2i \neq 3 - 2i$ because the numbers must be identical to be equal.

- **Operations:**
  - **Addition and Subtraction:**
    - Combine complex numbers as though the i were a variable.

      EX: $(4 + 5i) + (7 - 3i) = 11 + 2i$; $(-3 + 7i) - (5 - 8i) = -8 + 15i$
    - The sum or difference of complex numbers is another complex number. Even the number **21** is a complex number of the form $\mathbf{a + b}i$ where $\mathbf{a} = 21$ and $\mathbf{b} = 0$.
  - **Multiplication and Division:**
    - Multiply complex numbers using the methods for multiplying two binomials. Remember that $i^2 = -1$, so the answer is not complete until i^2 has been replaced with **-1** and simplified.

      EX: $(-3 + 5i)(1 - i) = -3 + 3i + 5i - 5i^2 = -3 + 3i + 5i - 5(-1) = -3 + 3i + 5i + 5 = 2 + 8i$

- Divide complex numbers by rationalizing the denominator. The answer is complete when there is no radical expression or ***i*** in the denominator and the answer is in simplest form.

  EX: The conjugate of the complex number **$-3 + 12i$** is **$-3-12i$**.

18 Graphing

Real Number Line

Chart of the graphs, on the real number line, of solutions to one-variable equations:

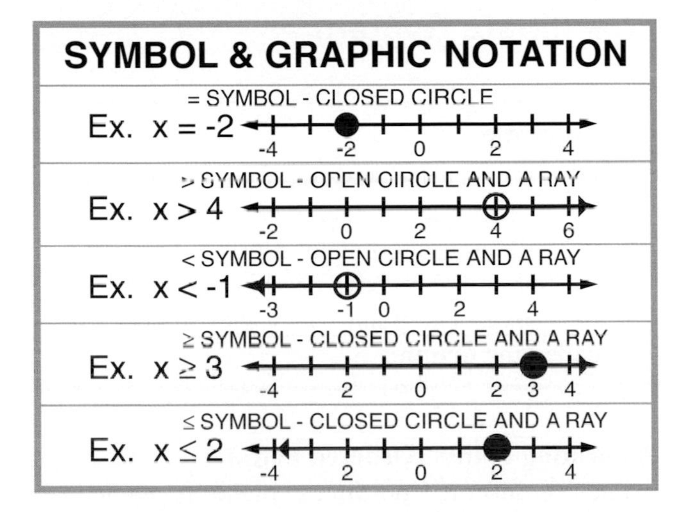

* Direction of ray is determined by picking (at random) a value on each side of the circle. Ray goes in direction of the point which makes the inequality true.

Absolute Value Statements

- **Equalities**: To solve $|\mathbf{ax+b}| = \mathbf{c}$, where $\mathbf{c} > \mathbf{0}$, solve both equations $\mathbf{ax} + \mathbf{b} = \mathbf{c}$ and $\mathbf{ax} + \mathbf{b} = \mathbf{-c}$, and graph the union of the two solutions.

Inequalities:

- To solve $|ax + b| < c$, where $c > 0$, solve $ax + b < c$ and $ax + b > -c$; these two inequalities may be written as one $-c < ax + b < c$; graph the intersection of the two solutions.
- To solve $|ax + b| > c$, where $c > 0$, solve $ax + b > c$ or $ax + b < -c$; graph the union of the two solutions.

Rectangular (Cartesian) Coordinate System

NOTES

Method, using two perpendicular lines (intersecting at 90 degree angles), for locating and naming points of a plane. The vertical line is the **y-axis**. The horizontal line is the **x-axis**. The point where they intersect is called the **origin**.

Locating Points (Ordered Pairs)

Each point on coordinate plane is named or located by using an ordered pair of numbers separated by a comma and enclosed in a set of parenthesis; first number is **x**-coordinate or **abscissa;** second number is **y**-coordinate or **ordinate**; that is, an ordered pair is of the **form (x,y).** The **origin** is **(0,0).**

Quadrants

The **x-axis** and the **y-axis** separate the plane into fourths. Each fourth is called a **quadrant.** The quadrants are labeled using Roman numerals, starting in the upper right section, and continuing counterclockwise through quadrants I, II, III, and IV (which is located in the lower right section).

Distance Formula: $d=\sqrt{(a-c)^2+(b-d)^2}$

Finds distance between two points, **(a,b)** and **(c,d)**; is derived from the application of the Pythagorean Theorem and always results in a non-negative number.

Midpoint Formula: $\left(\frac{x_1+x_2}{2}, \frac{y_1+y_2}{2}\right)$

Determines the coordinates of the midpoint of a line segment with endpoints of (x_1, y_1) and (x_2, y_2).

19 Lines

Slope of a Line

NOTES

The slope of a line can loosely be described as the slant of the line. If the line slants up on the right end of the line then the slope will be a positive number. If the line slants up on the left end of the line then the slope will be a negative number. If the line is horizontal then the slope is zero. If the line is vertical then the line has no slope, it is undefined.

- **Formula:** If line is not vertical then slope (indicated by **m**) can be found using two distinct points **A** = (x_1, y_1) and **B** = (x_2, y_2) of the line and using **x-coordinates** and **y-coordinates** in the formula:
$$m = \frac{y_2 - y_1}{x_2 - x_1} = \frac{\textbf{change in y}}{\textbf{change in x}} = \frac{\Delta y}{\Delta x} = \frac{\textbf{rise}}{\textbf{run}}$$
- **Parallel:** The slopes of parallel lines are equal.

- **Perpendicular:** The slopes of perpendicular lines are negative reciprocals. If the slope of $\mathbf{L_1}$ is $\mathbf{m_1}$ and the slope of $\mathbf{L_2}$ is $\mathbf{m_2}$, and the lines are perpendicular then $\mathbf{m_1} = \frac{\mathbf{-1}}{\mathbf{m_2}}$ or $\mathbf{(m_1)(m_2)} = -\mathbf{1}$.

  EX: If the slope of a line is $\frac{-1}{2}$ then the slope of the line which is perpendicular to it is **+2.**

Linear Equations (Equations of Lines)

- Since the coordinate system has an **x-axis** and a **y-axis,** lines which intersect the **x-axis** contain the variable **x** in the linear equation; lines which intersect the **y-axis** contain the variable **y** in the linear equation; and, lines which intersect both the **x-axis** and the **y-axis** have both variables **x** and **y** in the linear equation.
- **Slope-intercept form** of equation of a line is $\mathbf{y = mx + b}$ where **m** is the slope of the line and **b** is the **y-intercept (y-value** of the point where the line intersects **y-axis**).
- **Standard form** of the equation of a line is $\mathbf{ax+by=c}$ where the number values for **a, b**, and **c** are integers (note that the **b** does not represent the **y-intercept** in this form).

Graphing Lines

When equation of a line is known, it may be graphed in any of the following ways:

- **Horizontal lines** have equations which simplify to the form $\mathbf{y = b}$**,** where **b** is the **y-intercept.** The slope of these lines is zero.

- **Vertical lines** have equations which simplify to form **x = c,** where **c** is the **x-intercept.** They have no slope.
- **Find at least two points** which make the equation true and are therefore on the line. Finding a third point is one method of checking for errors. If all three points do not form a line then there is an error in at least one of the points. To find these points:
  - Choose a number at random.
  - Substitute the number into the linear equation for either the **x** or the **y** variable in the equation.
  - Solve the resulting equation for the other variables.
  - The randomly selected number and solution number result in one point: **(x, y)**.
  - Repeat above steps as indicated until the desired number of points have been created.
  - Plot points and connect them; resulting graph should be a line.
- **Plot the x-intercept and the y-intercept:**
  - Substitute zero for the **y** variable in the equation and solve for **x** to find the **x-intercept.**
  - Substitute zero for the **x** variable in the equation and solve for **y** to find the **y-intercept.**
  - Plot these two points and draw the graph of the line which contains them.
  - **Note:** Lines which have the same point as the **y-intercept** and the **x-intercept,** that is, the origin **(0,0)**, must have at least one other point located in order to draw the graph of the line.
- **Write the equation in the slope-intercept form,** plot the point where the line crosses the **y-axis** (the **b** value), use the slope to plot additional points on the line (rise over run). Connect the points to draw the graph of the line.

- **Find the slope of the line and one point on the line.** Plot the point first, then use the slope to plot additional points on the line. That is, count the slope as rise over run beginning at the point which was just plotted.

Finding the Equation of a Line

- **Horizontal Lines:** the slope is zero and the equation of the line takes the form of **y = b,** where **b** is the **y-intercept** (the **y**-value of the point of intersection of the line and the **y-axis**).
- **Vertical Lines:** there is no slope and the equation of the line takes the form of **x = c**, where **c** is the **x-intercept** (the **x**-value of the point of intersection of the line and the **x-axis**).
- **Neither Horizontal nor Vertical:**
  - ◆ Given the slope and the **y-intercept** values: substitute these numerical values in the slope-intercept form of a linear equation, **y = mx + b**, where **m** is the slope and **b** is the **y**-intercept.
  - ◆ **Given the slope and one point, either:**
    - Use the formula for slope, $m=\frac{y_2-y_1}{x_2-x_1}$, or the point-slope form $(x_2-x_1)m=(y_2-y_1)$.
      - > Substitute the coordinates from the point for the x_1 and y_1 variables and the slope value for the **m**.
      - > The equation is then changed to standard form **ax + by = c**, where **a**, **b**, and **c** are integers.
    - Or, use the slope-intercept form of linear equation, **y = mx + b**, twice.

> The first time substitute the coordinates from the point in the equation for the variables **x** and **y**, and substitute the slope value for the **m**; solve for **b**.

> The second time use the slope-intercept form of a linear equation. Substitute the numerical value for the slope **m** and the intercept **b**, leave the variables **x** and **y** in the equation. The result is the equation of the line in slope-intercept form.

◆ **Given two points:**

- Using the points in the slope formula, find the value for the slope, **m**.
- Using the slope value and either one of the two points (pick at random), follow the steps given above for the slope and one point.

◆ **Given the equation of another line:**

- Parallel to the requested line

> Use the given equation to find the slope. Parallel lines have the same slope.

> Use this slope value and any other given information and follow the steps above, depending on the type of information which is given.

- Perpendicular to the requested line

> Use the given equation to find the slope. The slope of the requested linear equation is the negative reciprocal of this slope, so change the sign and flip the number to find the slope of the requested line.

> Use this slope value and any other information given in steps above, depending on the type of information which is given.

Graphing Linear Inequalities

- **Graphs of linear inequalities such as $>$ and $<$ are half planes.**
  - Replace the inequality symbol with $=$ and graph this linear equality as a broken line to indicate that it is only the separation and not part of the graph.
  - To graph the inequality, randomly pick any point above this line and any point below this line.
  - Substitute each point into the original inequality.
  - Whichever point makes the inequality true is in the graph of the inequality so shade all points in the coordinate plane which are on the same side of the line with this point.
- **Graphs of linear inequalities such as $\geq$ and $\leq$ include both the half planes and the lines.**
  - The same methods given in items above apply except the line is drawn in solid form because it is part of the graph since the inequalities also include the equal sign.

Finding the Intersection of Lines: Systems of Linear Equations

NOTES

The purpose of finding the intersection of lines is to find the point which makes two or more equations true at the same time. These equations form a system of equations. These methods are extremely useful in solving word problems.

- **The system of equations is either:**
  - **Consistent;** that is, the lines intersect at one point.
  - **Inconsistent;** that is, the lines are parallel and since they do not intersect, there is no solution to the system of equations. The solution set is the empty set.
  - **Dependent;** that is, the graphs are the same line. All of the points which make one equation true also make the others true. The lines have all points in common and are therefore dependent equations.
- **To find the solution to a system of equations use one of these methods:**
  - **Graph Method** - Graph the equations and locate the point of intersection, if there is one. The point can be checked by substituting the **x** value and the **y** value into all of the equations. If it is the correct point it should make all of the equations true.

Note: *This method is weak, since an approximation of the coordinates of the point is often all that is possible.*

◆ **Substitution Method** for solving consistent systems of linear equations includes following steps:

- **Solve** one of the equations for one of the variables. It is easiest to solve for a variable which has a coefficient of one (if such a variable coefficient is in the system) because fractions can be avoided until the very end.
- **Substitute** the resulting expression for the variable into the **other** equation, not the same equation which was just used.
- **Solve** the resulting equation for the remaining variable. This should result in a numerical value for the variable, either **x** or **y,** if the system was originally only two equations.
- **Substitute** this numerical value back into one of the original equations and solve for the other variable.
- **The solution** is the point containing these **x-** and **y**-values, **(x,y)**.
- **Check the solution** in all of the original equations.

◆ **Elimination Method** or the **Add/Subtract Method** or the **Linear Combination Method** - Eliminate either the **x** or the **y** variable through either addition or subtraction of the two equations. These are the steps for consistent systems of two linear equations.

- **Write both equations in the same order**, usually $ax + by = c$, where **a, b**, and **c** are real numbers.
- **Observe the coefficients** of the **x** and **y** variables in both equations to determine:
  - > If the **x** coefficients or the **y** coefficients are the **same**, **subtract** the equations.
  - > If they are additive inverses (opposite signs: such as **3** and **-3**), **add** the equations.
  - > If the coefficients of the **x** variables are not the same and are not additive inverses, and the same is true of the coefficients of the **y** variables, then multiply the equations to make one of these conditions true so the equations can be either added or subtracted to eliminate one of the variables.
- The above steps should result in **one equation with only one variable, either x or y, but not both.** If the resulting equation has both **x** and **y**, an error was made in following the steps indicated above. Correct the error.
- **Solve** the resulting equation for the one variable (**x** or **y**).
- **Substitute** this numerical value back into either of the original equations and solve for the one remaining variable.
- **The solution** is the point (**x, y**) with the resulting **x**- and **y**-values.
- **Check the solution** in all of the original equations.

◆ **Matrix method**: Involves substantial matrix theory for a system of more than two equations and will not be covered here. Systems of two linear equations can be solved using **Cramer's Rule** which is based on determinants.

- For the system of equations: $\mathbf{a_1x + b_1y = c_1}$ and $\mathbf{a_2x + b_2y = c_2}$, where all of the **a, b**, and **c** values are real numbers, the point of intersection is $\mathbf{(x,y)}$ where $\mathbf{x = (D_x)/D}$ and $\mathbf{y=(D_y)/D}$.
- The determinant **D** in these equations is a numerical value found in this manner:

$$D = \begin{vmatrix} a_1 & b_1 \\ a_2 & b_2 \end{vmatrix} = a_1 b_2 - a_2 b_1$$

- The determinant $\mathbf{D_x}$ in these equations is a numerical value found in this manner:

$$D_x = \begin{vmatrix} c_1 & b_1 \\ c_2 & b_2 \end{vmatrix} = c_1 b_2 - c_2 b_1$$

- The determinant $\mathbf{D_y}$ in these equations is a numerical value found in this manner:

$$D_y = \begin{vmatrix} a_1 & c_1 \\ a_2 & c_2 \end{vmatrix} = a_1 c_2 - a_2 c_1$$

- Substitute the numerical values found from applying the determinant formulas into $x = \frac{(D_x)}{D}$ and $y = \frac{(D_y)}{D}$.

20 Functions

Basic Concepts

■ **Relation**

◆ Set of ordered pairs; in the coordinate plane, **(x,y)**.

- If a relation, **R,** is the set of ordered pairs **(x,y)** then the **inverse** of this relation is the set of ordered pairs **(y,x)** and is indicated by the notation $\mathbf{R^{-1}}$.

■ **Domain**

◆ Set of the first components of the ordered pairs of the relation; in the coordinate plane, a set of the **x**-values.

■ **Range**

◆ Set of the second components of the ordered pairs of the relation; in the coordinate plane, a set of the **y**-values.

■ **Function**

◆ Relation in which there is exactly one second component for each of the first components.

- **y** is a function of **x** if exactly one value of **y** can be found for each value of **x** in the domain; that is, each **x**-value has only one **y**-value but different **x**-values could have the same **y**-value, so the **y**-values may be used more than once for different **x**-values.

■ **Vertical Line Test**

◆ Indicates a relation is also a function if no vertical line intersects the graph of the relation in more than one point.

■ **One-to-One Functions**

◆ A function, **f**, is one-to-one if $\mathbf{f(a) = f(b)}$ only when $\mathbf{a = b}$

■ **Horizontal Line Test**

◆ Indicates a one-to-one function if no horizontal line intersects the graph of the function in more than one point.

Notation

■ **f(x) is read as "f of x"**

◆ Does not indicate the operation of multiplication. Rather, it indicates a function of **x**.

- $\mathbf{f(x)}$ is another way of writing $\mathbf{y}$ in that the equation $\mathbf{y = x + 5}$ may also be written as $\mathbf{f(x) = x + 5}$ and the ordered pair $\mathbf{(x,y)}$ may also be written $\mathbf{(x,f(x))}$.
- To evaluate $\mathbf{f(x)}$, use whatever expression is found in the set of parentheses following the **f** to substitute into the rest of the equation for the variable **x**, then simplify completely.

■ **Composite Functions: f [g(x)]**

◆ Composition of the function **f** with the function **g**, and it may also be written as $\mathbf{f \circ g(x)}$.

◆ The composition, $\mathbf{f[g(x)]}$, is simplified by evaluating the **g** function first and then using this result to evaluate the **f** function.

■ $\mathbf{(f + g)(x)}$ **equals** $\mathbf{f(x) + g(x)}$

That is, it represents the sum of the functions.

- **(f - g)(x) equals f(x) - g(x)**
  That is, it represents the difference of the functions.
- **(fg)(x) equals f(x) • g(x)**
  That is, it represents multiplication of the functions.
- **(f/g)(x) equals f(x) / g(x)**
  That is, it represents the division of **f(x)** by **g(x)**.

21 Types of Functions

NOTES

All linear equations, except those for vertical lines, are functions.

Polynomial Functions

- **Written Form**
  - $f(x) = a_n x^n + a_{n-1} x^{n-1} + a_{n-2} x^{n-2} + ... + a_1 x + a_0$ for real number values for all of the **a**'s, $a_n \neq 0$.
- **May have to have restricted domains and/or ranges to qualify as a function**
  - Without restrictions some equations would only qualify as relations and not functions.
- **Find the equation of the inverse of a function**
  - Exchange **x** and **y** variables in equation of the function and then solve for **y**. Finally replace **y** with $f^{-1}(x)$. Not all inverses of functions are also functions.
- **To Graph**
  - Use the Remainder Theorem, if a polynomial **P(x)** is divided by **x - r**, the remainder is **P(r)**, to determine remainders through substitution.
  - Use the Factor Theorem, if a polynomial **P(x)** has a factor **x - r** if and only if **P(r) = 0**, to find the zeros, roots, and factors of the polynomial.

- Find number of turning points of graph of a polynomial of degree **n** to be **n – 1** turning points at most.
- **Sketch,** using slashed lines, all vertical and/or horizontal asymptotes, if there are any.
- **Find the signs** of **P(x)** in intervals between and to each side of the intercepts. This is done to determine the placement of the graph above or below the **x-axis.**
- **Plot a few points** in each interval to find the exact graph placement. Also plot all intercepts.
- **Note:** The graphs of inverse functions are reflections about the graph of the linear equation $\mathbf{y = x}$.

Exponential Functions

Definition

- An **exponential function** has the form $\mathbf{f(x) = a^n}$, where $\mathbf{a > 0}$, $\mathbf{a \neq 1}$, and the constant real number, **a**, is called the **base**.

Properties

- The graph always intersects the **y-axis** at **(0,1)** because $\mathbf{a^0 = 1}$.
- The **domain** is the set of all real numbers.
- The **range** is the set of all positive real numbers because **a** is always positive.
- Inverses of exponential functions are **logarithmic functions.**

Logarithmic Functions

Definition

- A logarithm is an exponent such that for all positive numbers **a,** where $\mathbf{a \neq 1}$, $\mathbf{y = \log_a x}$ **if and only if** $\mathbf{x = a^y}$; notice that this is the logarithmic function of base **a**.

- The **common logarithm, log x,** has no base indicated and the understood base is always **10.**
- The **natural logarithm, ln x**, has no base indicated, is written **ln** instead of **log,** and the understood base is always the number **e.**

Properties with the variable *a* representing a positive real number not equal to one:

- $a^{\log_a x} = x$
- $\log_a a^x = x$
- $\log_a a = 1$
- $\log_a 1 = 0$
- If $\log_a u = \log_a v$, then $u = v$
- If $\log_a u = \log_b u$ and $u \neq 1$, then $a = b$.
- $\log_a xy = \log_a x + \log_a y$
- $\log_a\left(\frac{x}{y}\right) = \log_a x - \log_a y$
- $\log_a\left(\frac{1}{x}\right) = -\log_a x$
- $\log_a x^n = n(\log_a x)$, where **n** is a real number.
- Change of Base Rule: If $a > 0$, $a \neq 1$, $b > 0$, $b \neq 1$, and $x > 0$ then $\log_a x = \frac{(\log_b x)}{(\log_b a)}$
- Finding Natural Logarithms: $\ln x = \frac{(\log x)}{(\log e)}$

Common Mistakes!

- $\log_a(x+y) = \log_a x + \log_a y$ FALSE!
- $\log_a x^n = (\log_a x)^n$ FALSE!
- $\frac{(\log_a x)}{(\log_a y)} = \log_a(x-y)$ FALSE!

- **Solving Logarithmic Equations**
  - Put all logarithm expressions on one side of the equals sign.
  - Use the properties to simplify the equation to one logarithm statement on one side of the equals sign.
  - Convert the equation to the equivalent exponential form
  - Solve and check the solution

Rational Functions

NOTES

Definition: $f(x) = \frac{P(x)}{Q(x)}$ where $\mathbf{P(x)}$ and $\mathbf{Q(x)}$ are polynomials which are relatively prime (lowest terms), $\mathbf{Q(x)}$ has degree greater than zero, and $\mathbf{Q(x) \neq 0}$.

To Graph

- **Domain**
  - The domain is all real numbers except for those numbers which make $\mathbf{Q(x) = 0}$.
- **Intercepts**
  - **y-intercept:** Set $\mathbf{x = 0}$ and solve for $\mathbf{y}$. There is one $\mathbf{y}$-intercept. If $\mathbf{Q(x) = 0}$ when $\mathbf{x = 0}$ then $\mathbf{y}$ is undefined and the function does not intersect the $\mathbf{y}$-axis.
  - **x-intercepts:** Set $\mathbf{y = 0}$. Since $f(x) = \frac{P(x)}{Q(x)}$ can equal zero only when $\mathbf{P(x) = 0}$, the $\mathbf{x}$-intercepts are the roots of the equation $\mathbf{P(x) = 0}$.

Asymptotes

A line which the graph of the function approaches, getting closer with each point, but never intersects.

Horizontal Asymptotes

- Horizontal asymptotes exist when the degree of **P(x)** is less than or equal to the degree of **Q(x).**
- The **x**-axis is a horizontal asymptote whenever **P(x)** is a constant and has degree equal to zero.
- Steps to find horizontal asymptotes
  - Factor out the highest power of **x** found in **P(x)**.
  - Factor out the highest power of **x** found in **Q(x)**.
  - Reduce the function; that is, cancel common factors found in **P(x)** and **Q(x)**.
  - Let $|x|$ increase, and disregard all fractions in **P(x)** and in **Q(x)** which have any power of **x** greater than zero in the denominators; because these fractions approach zero and may be disregarded completely.
  - **When the result of the previous step is:**
    - > a constant, **c**, the equation of the horizontal asymptote is $y = c$.
    - > a fraction such as c/x^n where **c** is a constant and $n \neq 0$, the asymptote is the **x-axis**.
    - > neither a constant nor a fraction, there is no horizontal asymptote.

Vertical Asymptotes

- Vertical asymptotes exist for values of **x** which make $Q(x) = 0$; that is, for values of **x** which make the denominator equal to zero, and therefore make the rational expression undefined.

- ◆ There can be several vertical asymptotes.
- ◆ Steps to find vertical asymptotes:
  - Set the denominator, **Q(x)**, equal to zero
  - Factor if possible
  - Solve for **x**
  - The vertical asymptotes are vertical lines whose equations are of the form **x = r**, where **r** is a solution of **Q(x) = 0** because each **r** value will make the denominator, **Q(x)**, equal to zero when it is substituted for **x** into **Q(x)**.

Symmetry

■ **Description:**

- ◆ Graphs are symmetric with respect to a line if, when folded along the drawn line, and the two parts of the graph then land upon each other.
- ◆ Graphs are symmetric with respect to the origin if, when the paper is folded twice, the first fold being along the **x-axis** (do not open this fold before completing the second fold) and the second fold being along the **y-axis**, the two parts of the graph land upon each other.

■ **Graphs are symmetric with respect to:**

- ◆ The **x-axis** if replacing **y** with **-y** results in an equivalent equation;
- ◆ The **y-axis** if replacing **x** with **-x** results in an equivalent equation;
- ◆ The **origin** if replacing both **x** with **-x** and **y** with **-y** results in an equivalent equation.

■ **Determine Points**

- ◆ Create a few points, by substituting values for **x** and solving for **f(x)**, which make the rational function equation true.

- Include points from each region created by the vertical asymptotes (choose values for **x** from these regions).
- Include the **y-intercept** (if there is one) and any **x-intercepts.**
- Apply symmetry (if the graph is found to be symmetric after testing for symmetry) to find additional points; that is, if the graph is symmetric with respect to the **x-axis** and point **(3,-7)** makes the equation **f(x)** true, then the point **(-3,-7)** will be on the graph and should also make the equation true.

Plot the Graph

- Sketch any horizontal or vertical asymptotes by drawing them as broken or dashed lines.
- Plot the points, some from each region created by the vertical asymptotes, which make the equation **f(x)** true.
- Draw the graph of the rational function equation, **f(x) = P(x) / Q(x)**, applying any symmetry which applies.

22 Sequences & Series I

Definitions

- **Infinite Sequence:** is a function with a domain which is the set of positive integers; written as $\mathbf{a_1}$, $\mathbf{a_2}$, $\mathbf{a_3}$, **....** with each $\mathbf{a_i}$ representing a term.
- **Finite Sequence:** is a function with a domain of only the first **n** positive integers; written as $\mathbf{a_1}$, $\mathbf{a_2}$, $\mathbf{a_3}$, **...**, $\mathbf{a_{n-1}}$, $\mathbf{a_n}$
- **Summation:** $\sum_{k=1}^{n} a_k = a_1 + a_2 + ... + a_{n-1} + a_n$

  where **k** is the index of the summation and is always an integer which begins with the value found at the bottom of the summation sign and increases by 1 until it ends with the value written at the top of the summation sign.
- ***n*th Partial Sum:** $S_n = \sum_{k=1}^{n} a_k = a_1 + a_2 + ... + a_{n-1} + a_n$
- **Arithmetic Sequence or Arithmetic Progression:** is a sequence in which each term differs from the preceding term by a constant amount, called the **common difference;** that is, $\mathbf{a_n = a_{n-1} + d}$ where **d** is the **common difference.**

- **Geometric Sequence or Geometric Progression:** is a sequence in which each term is a constant multiple of the preceding term; that is, $\mathbf{a_n = ra_{n-1}}$ where **r** is the constant multiple and is called the common ratio.
- **n! = n(n - 1)(n - 2)(n - 3) ... (3)(2)(1);** this is read **"n factorial." Note: 0! = 1**

Properties of Sums, Sequences & Series

- $\sum_{k=1}^{n} (a_k \pm b_k) = \sum_{k=1}^{n} a_k \pm \sum_{k=1}^{n} b_k$
- $\sum_{k=1}^{n} ca_k = c\sum_{k=1}^{n} a$ where **c** is a constant
- $\sum_{k=1}^{n} c = nc$ where **c** is a constant
- The nth term of an arithmetic sequence is $\mathbf{a_n = a_1 + (n-1)d}$, where **d** is common difference.
- The sum of the first **n** terms of an arithmetic sequence with $\mathbf{a_1}$ as the first term and **d** as the common difference is $S_n = \frac{n}{2}(a_1 + a_n)$ OR $S_n = \frac{n}{2}[2a_1 + (n-1)d]$
- The nth term of a geometric sequence with $\mathbf{a_1}$ as the first term and r as the common ratio is $\mathbf{a_n = a_1 r^{n-1}}$.
- The sum of the first **n** terms of a geometric sequence with $\mathbf{a_1}$ as the first term and **r** as the common ratio and $\mathbf{r \neq 1}$ is $S_n = \frac{[a_1(1-r^n)]}{1-r}$

- The sum of the terms of an infinite geometric sequence with $\mathbf{a_1}$ as the first term and $\mathbf{r}$ as the common ratio where $|\mathbf{r}| < \mathbf{1}$, is $\frac{\mathbf{a_1}}{\mathbf{1-r}}$ if $|\mathbf{r}| > \mathbf{1}$ or $|\mathbf{r}| = \mathbf{1}$, the sum does not exist
- The $\mathbf{r}^{\text{th}}$ term of the binomial expansion of $\mathbf{(x + y)^n}$ is $\frac{\mathbf{n!}}{\mathbf{[n-(r-1)]!(r-1)!}}\mathbf{x^{n-(r-1)}y^{(r-1)}}$

23 Conic Sections

NOTES

Below are general equation forms of conic sections. These general forms can be used both to graph and to determine equations of conic sections. The values for **h** and **k** can be any real number, including zero.

Description

Conic sections represent the intersections of a plane and a right circular cone; that is, parabolas, circles, ellipses, and hyperbolas. In addition, when the plane passes through the vertex of the cone it may determine a degenerate conic section; that is, a point, line, or two intersecting lines.

General Equation

The general form of the equation of a conic section with axes parallel to the coordinate axes is: $\mathbf{Ax^2 + Bxy + Cy^2 + Dx + Ey + F = 0}$ where **A** and **C** are not both zero.

Type: Line

- **General Equation: $y = mx + b$**
  - **Notation:**
    - **m** is slope
    - **b** is **y-intercept**
  - **Values:**
    - **$m > 0$** then the line is higher on the right end.
    - **$m < 0$** then the line is higher on the left end

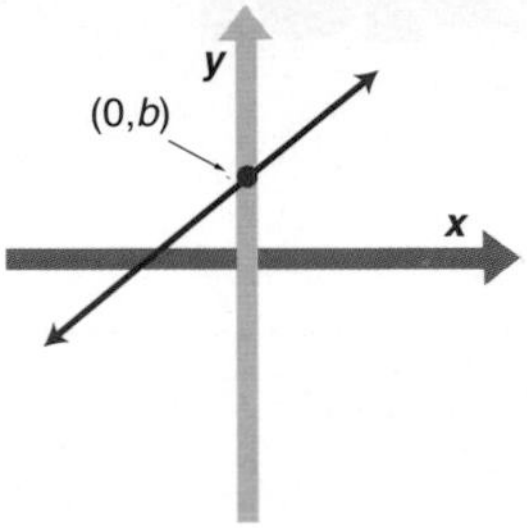

Type: Horizontal Line

- **General Equation: $y = b$**
  - **Notation:**
    - **b** is **y-intercept**
  - **Values:**
    - **$m = 0$** then the line is horizontal through **(0, b)**

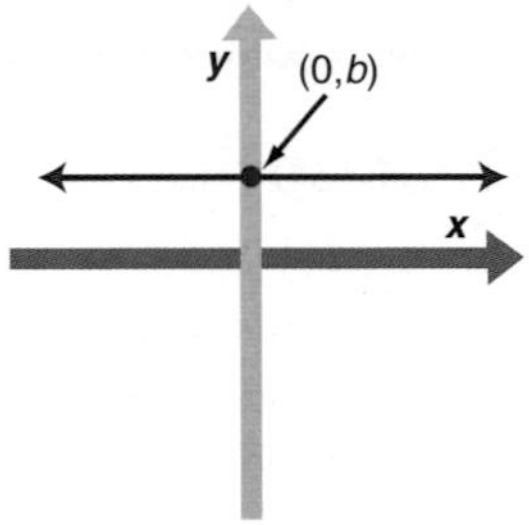

Type: Vertical Line

- **General Equation: $x = c$**
  - **Notation:**
    - **c** is **x-intercept**
  - **Values:**
    - no slope
    - vertical line through **(c, 0)**

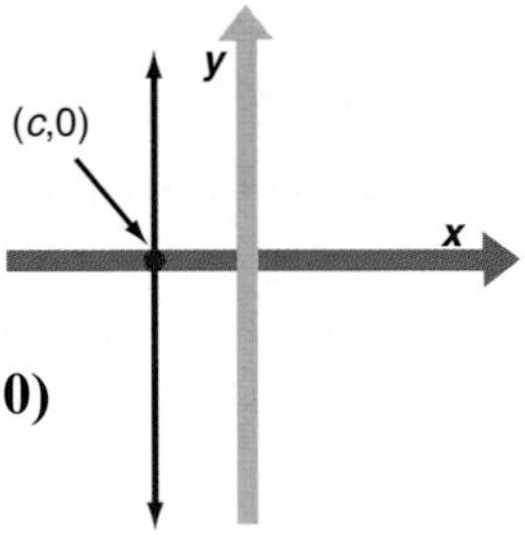

Type: Parabola

- **General Equation: $y = a(x - h)^2 + k$**
- **Standard Form: $(x - h)^2 = 4p(y - k)$**
  - **Notation:**
    - x^2 term and y^1 term
    - **(h, k)** is vertex
    - **(h, k ± p)** is center of focus where $p = \frac{1}{4a}$
    - **y = k ± p** is directrix equation where $p = \frac{1}{4a}$
  - **Values:**
    - **a > 0** then opens up.
    - **a < 0** then opens down
    - **x = h** is equation of line of symmetry
    - larger |**a**| = thinner parabola; smaller |**a**| = fatter parabola

Type: Parabola

- **General Equation: $x = a(y - k)^2 + h$**
- **Standard Form: $(y - k)^2 = 4p(x - h)$**
  - **Notation:**
    - x^1 term and y^2 term
    - **(h, k)** is vertex
    - **(h ± p, k)** is focus where $p = \frac{1}{4a}$
    - **x = h ± p** is directrix equation where $p = \frac{1}{4a}$

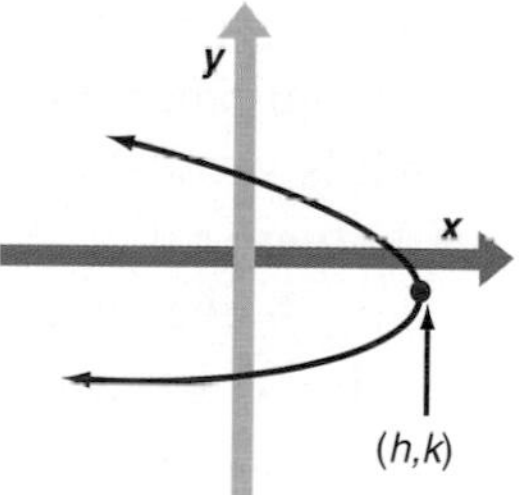

  - **Values:**
    - **a > 0** then opens right
    - **a < 0** then opens left
    - **y = k** is equation of line of symmetry

Type: Circle

■ **General Equation:**

$(\mathbf{x} - \mathbf{h})^2 + (\mathbf{y} - \mathbf{k})^2 = \mathbf{r}^2$

◆ **Notation:**

- $\mathbf{x}^2$ term *and* $\mathbf{y}^2$ term both with the same positive coefficient
- $\mathbf{r}^2$ is a positive number
- **(h, k)** is center
- **r** is radius

◆ **Values:** none

Type: Ellipse

■ **General Equation:**

$$\frac{(\mathbf{x}-\mathbf{h})^2}{\mathbf{a}^2} + \frac{(\mathbf{y}-\mathbf{k})^2}{\mathbf{b}^2} = 1$$

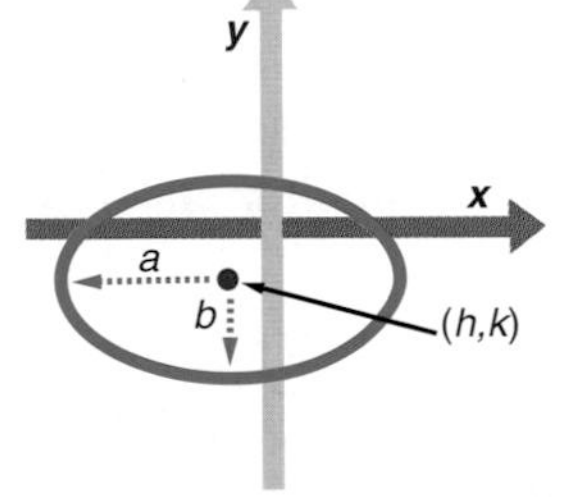

◆ **Notation:**

- $\mathbf{x}^2$ term and $\mathbf{y}^2$ term with different coefficients
- **(h, k)** is center
- **a** is horizontal distance to left and right of **(h, k)**
- **b** is vertical distance above and below **(h, k)**

◆ **Values:**

- $\mathbf{a} > \mathbf{b}$, then major axis is horizontal and foci are $(\mathbf{h} \pm \mathbf{c}, \mathbf{k})$, where $\mathbf{c}^2 = \mathbf{a}^2 - \mathbf{b}^2$
- $\mathbf{b} > \mathbf{a}$, then major axis is vertical and foci are $(\mathbf{h}, \mathbf{k} \pm \mathbf{c})$ where $\mathbf{c}^2 = \mathbf{b}^2 - \mathbf{a}^2$

Type: Hyperbola

■ **General Equation:**

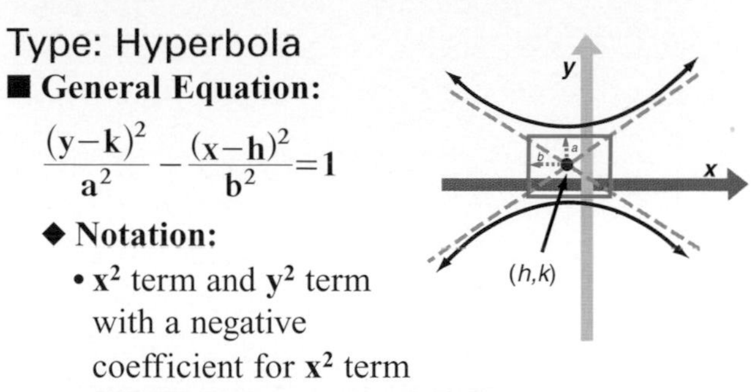

$$\frac{(y-k)^2}{a^2} - \frac{(x-h)^2}{b^2} = 1$$

◆ **Notation:**

- x^2 term and y^2 term with a negative coefficient for x^2 term
- (h, k) is center of a rectangle
- b is horizontal distance to left and right of (h, k)
- a is vertical distance above and below (h, k) to the vertices

◆ **Values:**

Equations of asymptotes $y - k = \pm \frac{a}{b}(x-h)$

Type: Hyperbola

■ **General Equation:**

$$\frac{(x-h)^2}{a^2} - \frac{(y-k)^2}{b^2} = 1$$

◆ **Notation:**

- x^2 term and y^2 term with a negative coefficient for y^2 term
- (h, k) is center of a rectangle
- a is horizontal distance to left and right of (h, k) to the vertices
- b is vertical distance above and below (h, k)

◆ **Values:**

Equations of asymptotes $y - k = \pm \frac{b}{a}(x-h)$

24 Problem Solving

Directions

- Read the problem carefully.
- Note the given information, the question asked and the value requested.
- Categorize the given information, removing unnecessary information.
- Read the problem again to check for accuracy, to determine what, if any, formulas are needed and to establish the needed variables.
- Write the needed equation(s) and determine the method of solution to use; this will depend on the degree of the equations, the number of variables and the number of equations.
- Solve the problem. Check the solution. Read the problem again to make sure the answer given is the one requested.

Odd Numbers, Even Numbers, Multiples

■ **Notation**

- **d** is the common difference between any two consecutive numbers of a set of numbers.

■ **Formulas**

- First number = **x**
- Second number = **x+d**
- Third number = **x+2d**
- Fourth number = **x+3d;** etc.

Ex: The first **5** multiples of **3** are **x, x+3, x+6, x+9**, and **x+12** because **d = 3**

Rectangles

■ **Notation**

- **P** is perimeter; **l** is length; **w** is width; **A** is area

■ **Formulas**

- $P = 2l + 2w$
- $A = lw$

Ex: The length of a rectangle is **5** more than the width and the perimeter is **38**.

Equation: $38 = 2(w + 5) + 2w$

Triangles

■ **Notation**

- **P** is perimeter;
- **S** is side length;
- **A** is area;
- **a** is altitude;
- **b** is base

NOTES

Altitude and base must be perpendicular (i.e., form 90° angles.)

■ **Formulas**

- ◆ $P = S_1 + S_2 + S_3$
- ◆ $A = \frac{1}{2}ab$

Ex: The base of a triangle is **3** times the altitude and the area is **24**.

Equation: $24 = \frac{1}{2} \cdot a \cdot 3a$

Circles

■ **Notation**

- ◆ **C** is circumference;
- ◆ **A** is area;
- ◆ **d** is diameter;
- ◆ **r** is radius;
- ◆ π is pi – 3.14...

■ **Formulas**

- ◆ $C = \pi d$
- ◆ $A = \pi r^2$
- ◆ $d = 2r$

Ex: The radius of a circle is 4 and the circumference is 25.12.

Equation: $25.12 = \pi \cdot 8$

Pythagorean Theorem

■ **Notation**

- ◆ **a** is a leg;
- ◆ **b** is a leg;
- ◆ **c** is a hypotenuse

NOTES

Hypotenuse is the longest side.

■ **Formulas**

◆ $a^2 + b^2 = c^2$

• **NOTE:** Applies to right triangles only.

Ex: The hypotenuse of a right triangle is 2 times the shortest leg. The other leg is $\sqrt{3}$ times the shortest leg.

Equation: $a^2 + (\sqrt{3}a)^2 = (2a)^2$

Money, Coins, Bills, Purchases

■ **Notation**

◇ **V** is currency value;

◇ **C** is number of coins, bills, or purchased items

■ **Formulas**

◇ $V_1C_1 + V_2C_2 = V_{total}$

Ex: Jack bought black pens at $1.25 each and blue pens at $0.90 each. He bought 5 more blue pens than black pens and spent $36.75.

Equation: $1.25x + 0.90(x+5) = 36.75$

Mixture

■ **Notation**

◇ V_1 is first volume;

◇ P_1 is first percent solution;

◇ V_2 is second volume;

◇ P_2 is second percent solution;

◇ V_F is final volume;

◆ P_F is final percent solution

NOTES

Water could be 0% solution and pure solution could be 100%.

■ **Formulas**

◆ $V_1P_1 + V_2P_2 = V_FP_F$

Ex: How much water should be added to 20 liters of 80% acid solution to yield 70% acid solution?

Equation: $x(0) + 20(0.80) = (x+20)(0.70)$

Work

■ **Notation**

◆ W_1 is rate of one person or machine multiplied by the time it would take for the entire job to be completed by 2 or more people or machines;

◆ W_2 is the rate of the second person or machine multiplied by time for entire job;

◆ **1** represents the whole job.

NOTES

Rate is the part of the job completed by one person or machine.

■ **Formulas**

◆ $W_1 + W_2 = 1$

Ex: John can paint a house in 4 days, while Sam takes 5 days. How long would they take if they worked together?

Equation: $\frac{1}{4}x + \frac{1}{5}x = 1$

Distance

■ **Notation**

- **d** is distance;
- **r** is rate; i.e. speed;
- **t** is time; value indicated in the speed, i.e. miles per hour has time in hours

NOTES

Add or subtract speed of wind or water current with the rate; (**r ± wind**) or (**r ± current**).

■ **Formulas**

- **d = rt**

  Ex: John traveled 200 miles in 4 hours.

  Equation: $200 = r \cdot 4$

- $d_{to} = d_{returning}$

  Ex: With a 30 mph head wind a plane can fly a certain distance in 6 hours. Returning, flying in opposite direction, it takes one hour less.

  Equation: $(r - 30)6 = (r + 30)5$

- $d_1 + d_2 = d_{total}$

  Ex: Lucy and Carol live 400 miles apart. They agree to meet at a shopping mall located between their homes. Lucy drove at 60mph, and Carol drove at 50mph and left one hour later.

  Equation: $60t + 50(t-1) = 400$

Simple Interest

Notation

- **I** is interest;
- **P** is principal, amount borrowed, saved, or loaned;
- **S** is total amount, or **I + P;**
- **r** is % interest rate;
- **t** is time expressed in years;
- **p** is monthly payment

Formulas

- **I = Prt**

  Ex: Anna borrowed $800 for 2 years and paid $120 interest.

  Equation: 120 = 800 r(2)

- **S = P + Prt**

  Ex: Alex borrowed $4600 at 9.3% for 6 months.

  Equation: S = 4600 + 4600 (.093)(.5)

  • **NOTE:** 9.3% = .093 and 6 months = 0.5 year

- $p = \frac{P+Prt}{t \cdot 12}$

  Ex: Evan borrowed $3,000 for a used car and is paying it off monthly over 2 years at 10% interest.

  Equation: p = [3,000 + 3000 (.1)(2)] / (2)(12)

Proportion & Variation

Notation

- **a, b, c, d,** are quantities specified in the problem; **$k \neq 0$.**

Formulas

- **Proportion:** $\frac{a}{b}=\frac{c}{d}$; cross multiply to get **ad = bc**

Ex: If 360 acres are divided between John and Bobbie in the ratio 4 to 5, how many acres does each receive?

Equation: $\frac{\text{John}}{\text{Bobbie}}$ so $\frac{4}{5}=\frac{J}{360-J}$

- **Direct Variation: $y = kx$**

Ex: If the price of gold varies directly as the square of its mass, and 4.2 grams of gold is worth $88.20, what will be the value of **10** grams of gold?

Equation: $88.20 = k(4.2)^2$; solve to find **$k = 5$**; then use the equation **$y = 5(10)^2$** where **y** is the value of **10** grams of gold.

- **Inverse variation:** $y=\frac{k}{x}$

Ex: If **a** varies inversely as **b** and **a = 4** when **b = 10,** find **a** when **b = 5.**

Equation: $4=\frac{k}{10}$ so **k = 40**; then $a=\frac{40}{5}$ to find **a**.

25 Real Numbers, Complex Numbers, Equality & Inequality

Properties of Real Numbers

For any real numbers a, b & c:

- **Closure**
  - For addition: $a + b$ is a real number
  - For multiplication: $a \cdot b$ is a real number
- **Commutative Property**
  - For addition: $a + b = b + a$
  - For multiplication: $ab = ba$
- **Associative Property**
  - For addition: $a + (b + c) = (a + b) + c$
  - For multiplication: $a(bc) = (ab)c$
- **Identity Property**
  - For addition: $a + 0 = a$ and $0 + a = a$
  - For multiplication: $a \cdot 1 = a$ and $1 \cdot a = a$
- **Inverse Property**
  - For addition: $a + (-a) = 0$ and $(-a) + a = 0$
  - For multiplication: $a \cdot \frac{1}{a} = 1$ and $\frac{1}{a} \cdot a = 1$
- **Distributive Property**
  - $a(b + c) = ab + ac$; $a(b - c) = ab - ac$; and $ab + ac = a(b + c)$; $ab - ac = a(b - c)$
- **Multiplication Property of Zero**

  $0 \cdot a = 0$ and $a \cdot 0 = 0$
- **Double Negative Property**

  $-(-a) = a$ or $-1(-1 \cdot a) = a$

Operations of Real Numbers

- **Absolute Value**
  - $|x| = x$ **if** $x \geq 0$ and $-x$ **if** $x < 0$. It is always a positive numerical value.
  - $|x| = |-x|$
  - $|x| \geq 0$
  - $|x - y| = |y - x|$
- **Addition:** If the **signs** of the numbers are the **same, add**. If the **signs** of the numbers are **different**, **subtract**. In both cases, the answer has the sign of the number with the larger absolute value.
- **Subtraction:** Change subtraction to **addition of the opposite number;** then follow the addition rules.
- **Multiplication:** Multiply the numbers then determine the sign of the answer. **Remember:** negative • negative = positive; positive • positive = positive; negative • positive = negative; positive • negative = negative; if the signs are the **same** the answer is **positive,** if signs are **different** the answer is **negative**.
- **Division:** Divide the numbers, then determine the sign of the answer using the same sign rules that apply to multiplication.

Operations of Complex Numbers

- **Definition:** $a \pm bi$ where **a, b** = Real numbers and $i = \sqrt{-1}$
- **Addition:** $(a \pm bi) + (c \pm di) = (a + c) + (b \pm d)i$
- **Subtraction:** $(a \pm bi) - (c \pm di) = (a - c) \pm (b - d)i$

- **Multiplication:** $(a \pm bi)(c \pm di) = (ac \mp bd) \pm (ad \pm bc)i$
- **Division:**

$$\frac{a \pm bi}{c \pm di} = \frac{a \pm bi}{c \pm di} \cdot \frac{c \mp di}{c \mp di} = \frac{(ac \mp bd) \pm (ad \pm bc)i}{c^2 \pm d^2}$$

Properties of Equality

For any real numbers a, b, and c:

- **Reflexive: a = a**
- **Symmetric:** If **a = b** then **b = a.**
- **Transitive:** If **a = b** and **b = c** then **a = c.**
- **Addition Property:** If **a = b** then **a + c = b + c** for any value of **c**.
- **Multiplication Property:** If **a = b** then **ac = bc** for any value of **c**.
- **Proportion Property:** If $\frac{a}{b} = \frac{c}{d}$, then **ad = bc**;

$$\frac{a}{c} = \frac{b}{d}; \frac{b}{a} = \frac{d}{c}; \frac{c}{a} = \frac{d}{b}; \frac{a \pm b}{b} = \frac{c \pm d}{d}; \frac{a-b}{a+b} = \frac{c-d}{c+d}.$$

Properties of Inequality

For any real numbers a, b & c:

- **Trichotomy:** Either **a = b, a < b,** or **a > b**.
- **Transitive:** If **a < b** and **b < c** then **a < c**; also, if **a > b** and **b > c** then **a > c**.
- **Addition Property:** If **a < b** then **a + c < b + c**; also, if **a > b** then **a + c > b + c** for any value of **c**.
- **Multiplication Property:** If **c > 0** and **a < b** then **ac < bc.** If **c > 0** and **a > b** then **ac > bc.** If **c < 0** and **a < b** then **ac > bc.** If **c < 0** and **a > b** then **ac < bc.** If **c = 0** then **ac = bc = 0**.

26 Review of Solving Equations—Steps in Detail

First Degree with One Variable

- Simplify the left side of the equals sign.
- Simplify the right side of the equals sign.
- Apply inverse operations until the variable is isolated.
  Note: If the statement is an inequality and multiplication or division by a negative number was used to distribute throughout the entire inequality, then the inequality symbol must be reversed to keep the statement true and the solution correct.

Systems of Equations

- A **consistent** system has one or more solutions.
- An **inconsistent** system has no solution.
- Methods of solution
  - Linear combination or elimination
    - Put both equations in standard form (alphabetical order with the constant on the right side of the equals sign).
    - Distribute through one or both equations so the coefficients (numbers in front of the variables) of one of the variables in both equations are opposite values (same digits with opposite signs).
    - Add the two equations (the result will be one equation with one variable).
    - Solve the resulting equation for the remaining variable.

- Put the numerical value for that variable back into one of the original equations and solve for the other variable.

◆ **Substitution**

- Solve one of the equations for one of the variables.
- Put the resulting statement into the other equation in place of the variable (the resulting equation will have only one variable).
- Solve the equation for the numerical value of the variable.
- Put this numerical value into either of the two beginning equations and solve for the second variable numerical value.

◆ **Graphing**

- Graph each equation on the coordinate plane.
- Find and label the point(s) of intersection, if there are any.

◆ **Cramer's Rule**

- Put both equations in standard form (alphabetical order with the constant on the right side of the equals sign): **ax + by = c** and **dx + ey = f.**
- Take the coefficients (numbers in front of the variables) and make a determinate of the system of equations, $D = \begin{vmatrix} a & b \\ d & e \end{vmatrix}$.
- Take the determinate of the system that was just created and replace the coefficients of one variable with the constants of the system (this is the determinate of that variable), $D_x = \begin{vmatrix} c & b \\ f & e \end{vmatrix}$.

- Take the original determinate of the system and replace the coefficients of the other variable with the constants of the system (this is the determinate of the other variable), $D_y = \begin{vmatrix} a & c \\ d & f \end{vmatrix}$.
- Solve each determinate by finding the difference of the cross product, $\begin{vmatrix} a & b \\ c & d \end{vmatrix} = ad - cb$.
- The solutions are $x = \frac{D_x}{D}$; and $y = \frac{D_y}{D}$.

Matrices & Linear Systems

- A matrix is a rectangular array of real numbers, called entries or elements, enclosed within brackets,

$$A = \begin{bmatrix} a_{11} & a_{12} & \cdots & a_{1n} \\ a_{21} & a_{22} & \cdots & a_{2n} \\ \vdots & \vdots & \vdots & \vdots \\ a_{m1} & a_{m2} & \cdots & a_{mn} \end{bmatrix},$$ where **n** = number of columns and **m** = number of rows.
- The **dimension** is "**m** by **n**".
- A **coefficient matrix** is formed by the coefficients of one variable of a system of linear equations forming each column of the matrix.
- An **augmented matrix** includes the constants of a system of linear equations separated by a vertical dashed line in the matrix.
- Row operations that transform an augmented matrix into an equivalent system:
  - Interchange any two rows.
  - Multiply every element of any row by a constant, **c**, where $c \neq 0$.

- Replace every element of any row by the sum of itself and a corresponding element of any other row.
- **Gauss-Jordan elimination** applies the row operations to an augmented matrix until all elements of the matrix, except the constants, are zeros and ones, with the ones forming a diagonal from upper left to lower right. The solutions of the system are then the constants that correspond to the coefficient values that are the ones; so $\left[\begin{array}{ccc|c} 1 & 0 & 0 & c_1 \\ 0 & 1 & 0 & c_2 \\ 0 & 0 & 1 & c_3 \end{array}\right]$ means $\mathbf{x} = \mathbf{c_1}$, $\mathbf{y} = \mathbf{c_2}$, and $\mathbf{z} = \mathbf{c_3}$.

- **Equal matrices** have the same dimensions and equal corresponding elements.
- The **sum** of two matrices is found by adding the corresponding elements of the matrices.
- **Scalar multiplication** is accomplished by multiplying every element of the matrix by the scalar ***c*.**
- The **difference** of two matrices is obtained by subtracting every corresponding element of the matrices.
- The **product** of two matrices can be found only if the number of columns of one matrix equals the number of rows of the other, and the number of rows of one equals the number of columns of the other; then, the elements of each row of one are multiplied by the elements of the columns of the other matrix, and these products are added to result in one number that is one element in the product matrix.
- A **determinant** is a number calculated from a square matrix (matrix with the same number of rows and columns), like $\begin{vmatrix} \mathbf{a} & \mathbf{b} \\ \mathbf{c} & \mathbf{d} \end{vmatrix} = \mathbf{ad - bc}$.

- The **cofactor** of the element $\mathbf{a}_{ij}$ is the minor of the element $\mathbf{a}_{ij}$ multiplied by $(\mathbf{-1})^{i+j}$.
- **Expansion by cofactors** is evaluating a determinate by forming the sum of the products obtained by multiplying each element of any row or any column by its cofactor.
- Cramer's Rule is discussed earlier in Systems of Equations.

Second Degree – Quadratic & Polynomial Equations

- **One Variable**
  - Methods
    - Factoring
      - Use inverse operations to set the equation equal to zero
      - Factor
      - Set each factor equal to zero
      - Solve each resulting equation
    - Quadratic Formula
      - Use inverse operations to set the equation equal to zero
      - Apply the quadratic formula, $x = \frac{-b \pm \sqrt{b^2 - 4ac}}{2a}$ where a, b, and c are numbers found in the equation that you just set equal to zero, $\mathbf{ax^2 + bx + c = 0}$; notice that the a is the coefficient of the squared term, b is the coefficient of the first degree term, and c is the constant without a variable.

Polynomials

- **Fundamental Theorem of Algebra:** Every polynomial of degree $n \geq 1$ has at least one root among the complex numbers; it has exactly n roots among the complex numbers when a root that is repeated k times is counted k times.

- **Conjugate Roots Theorem:** If $a + bi$, $b \neq 0$, is a root of a polynomial of degree $n \geq 1$ with real coefficients, then $a - bi$ is also a root.

- **Descartes's Rule of Signs:** If $P(x)$ is a polynomial with real coefficients, then:
    - The number of positive roots either is equal to the number of variations in sign of $P(x)$ or is less than the number of variations in sign by an even number, and
    - The number of negative roots either is equal to the number of variations in sign of $P(-x)$ or is less than the number of variations in sign by an even number.

Rational Equations

- Simplify the rational expressions on both sides of the equals signs by getting like denominators and adding $\frac{1}{a}+\frac{1}{b}=\frac{1}{c}+\frac{1}{d}$; becomes $\frac{b}{ab}+\frac{a}{ab}=\frac{d}{cd}+\frac{c}{cd}$ then $\frac{b+a}{ab}=\frac{d+c}{cd}$.

- Multiply the entire equation by the common denominators to eliminate them, getting $cd(b+a) = ab(d+c)$.

- Solve the resulting equation.

Radical Equations

- Isolate the radical expression on one side of the equation, if possible (if not possible, put the radical expressions on two opposite sides of the equals sign).
- Square, cube, or raise both entire sides of the equation to the needed power (the power that matches the index of the radical).
- Repeat steps 1 and 2 as needed until all radical expressions are gone.
- Solve the resulting equation.
- Check the solution(s) back in the original equation because there is a possibility of extraneous roots (solutions that do not make the original equation true).

27 Coordinate Plane

Review of Equations

- The ***x*-axis** is horizontal; the ***y*-axis** is vertical.
- Each point is named by an **ordered pair**, *(x, y)*.
- The distance between two points is

  $$d = \sqrt{(x_2 - x_1)^2 + (y_2 - y_1)^2}$$

- The midpoint of a line segment with endpoints (x_1, y_1) and (x_2, y_2) is $P(x, y)$ where $x = \frac{x_1 + x_2}{2}$ and $y = \frac{y_1 + y_2}{2}$.
- A **relation** is a set of ordered pairs.
- A **function** is a relation that has no x-values that are the same.
  - $f(x)$ is read "*f of x*" or "***the function of x***"
  - $(f + g)(x) = f(x) + g(x)$
  - $(f - g)(x) = f(x) - g(x)$
  - $(fg)(x) = f(x) \cdot g(x)$
  - $(f/g)(x) = f(x) / g(x)$
  - $g[f(x)] = g$ of $f(x)$
- **Vertical line** test of a function: A graph represents a function if and only if no vertical line intersects the graph in more than one point.

Polynomials

- Form: $f(x) = a_n x^n + a_{n-1} x^{n-1} + \ldots + a_1 x + a_0$ for real numbers, **a**, with $a_n \neq 0$.

- Restrictions on the coordinate values may be necessary for the polynomial to be a function.
- **Intermediate Value Theorem:** If $a < b$ and $f(x)$ is a polynomial function such that $f(a) \neq f(b)$, then f takes every value between $f(a)$ and $f(b)$ in the interval $[\mathbf{a}, \mathbf{b}]$.
- The graph of a polynomial function of degree n has at most $n - 1$ **turning points**.
- Inverse functions, $f^{-1}(x)$, are found by exchanging the x and the y variables in the equation; $f^{-1}[f(x)] = x$ for every x in the domain, and $f[f^{-1}(y)] = y$ for every y in the range.
- **Remainder Theorem:** If the polynomial $P(x)$ is divided by x - $\mathbf{r}$, the remainder is $P(\mathbf{r})$.
- **Factor Theorem:** The polynomial $P(x)$ has a factor x - $\mathbf{r}$ if and only if $P(\mathbf{r}) = \mathbf{0}$ and $\mathbf{r}$ is a root.
- If the coefficients of the polynomial $P(x) = \mathbf{a}_n\mathbf{x}^n + \mathbf{a}_{n-1}\mathbf{x}^{n-1} + \ldots + \mathbf{a}_1\mathbf{x} + \mathbf{a}_0$ are integers and $\frac{\mathbf{p}}{\mathbf{q}}$ is a rational root in lowest terms, then p is a factor of the constant term a_0, and q is a factor of the leading coefficient a_n.
- The graph of the rational function $f(x) = \frac{P(x)}{Q(x)}$ has a horizontal asymptote if the degree of $\mathbf{P(x)} \leq$ the degree of $\mathbf{Q(x)}$; and it has a vertical asymptote at $x = r$, if r is a real root of $\mathbf{Q(x)}$ but not of $\mathbf{P(x)}$.
- The tests for **symmetry** of graphs symmetric with respect to the ...
    - x-axis if replacing y in the equation with $-y$ results in an equivalent equation.
    - y-axis if replacing x in the equation with $-x$ results in an equivalent equation.
    - origin if replacing both x with $-x$ and y with $-y$ in the equation results in an equivalent equation.

Exponential Functions

- Form: $f(x) = a^n$, where $a > 0$, $a \neq 1$.
- Inverses of exponential functions are logarithmic functions.

Logarithmic Functions

- For all positive numbers a, where $a \neq 1$, $y = \log_a x$ if and only if $x = a^y$.
- The **common logarithm**, $\log x$, has a base of **10**, so $a = 10$ in the definition of $\log$.
- The **natural logarithm, $\ln x$**, has a base equal to the number e, $a = e \approx 2.71828$, in the definition of $\log$.
- Properties with $a > 0$ and $a \neq 1$
    - $a^{\log_a x} = x$
    - $\log_a a^x = x$
    - $\log_a a = 1$
    - $\log_a 1 = 0$
    - If $\log_a u = \log_a v$, then $u = v$.
    - If $\log_a u = \log_b u$ and $u \neq 1$, then $a = b$.
    - $\log_a xy = \log_a x + \log_a y$
    - $\log_a\left(\frac{x}{y}\right) = \log_a x - \log_a y$
    - $\log_a\left(\frac{1}{x}\right) = -\log_a x$
    - $\log_a x^n = n(\log_a x)$, where **n** is a real number
    - Change of base rule: if $a > 0$, $a \neq 1$, $b > 0$, $b \neq 1$, and $x > 0$ then $\log_b x = \frac{(\log_a x)}{(\log_a b)}$.
    - $\ln x = \frac{(\log x)}{(\log e)}$
    - **A one-to-one function** is a function that has no y-values that are the same.

Conic Sections

The general form of the equation of a conic section with axes parallel to the coordinate axes is $Ax^2 + Bxy + Cy^2 + Dx + Ey + F = 0$, where A and C are not both zero.

Lines

- Slope $= m = \frac{y_1 - y_2}{x_1 - x_2} = \frac{\Delta y}{\Delta x} = \frac{\text{rise}}{\text{run}}$
- Point-slope form of a linear equation: $y - y_1 = m(x - x_1)$.
- Slope-intercept form of a linear equation: $y = mx + b$, where m is the slope of the line and b is the y-intercept (point where the line crosses the y-axis).
- Standard form of a linear equation: $ax + by = c$, where a, b, and c are integers.
- Vertical lines have the equation $x = a$, where a is the x-value where the line crosses the x-axis.
- Horizontal lines have the equation $y = a$, where a is the y-value where the line crosses the y-axis.
- Two lines are parallel if and only if their slopes are equal; $m_1 = m_2$.
- Two lines are perpendicular if and only if their slopes are negative reciprocals; and $m_2 = -\frac{1}{m_1}$ and $m_1 = -\frac{1}{m_2}$.
- If $m > 0$, the line is increasing.
- If $m < 0$, the line is decreasing.
- If $m = 0$, the line is horizontal.
- If m is undefined, the line is vertical.

Parabolas

- General equation: $y = a(x - h)^2 + k$ (opens up/down); or $x = a(y - k)^2 + h$ (opens left/right)
- Standard form: $(x - h)^2 = 4p(y - k)$ (opens up/down); or $(y - k)^2 = 4p(x - h)$ (opens left/right)
- Where (h, k) is the vertex, $(h, k \pm p)$ or $(h \pm p, k)$ is the vertex with $p = \frac{1}{4a}$, $y = k \pm p$ or $x = h \pm p$ is the directrix, and $x = h$ or $y = k$ is the line of symmetry.

Circles

- General equation: $(x - h)^2 + (y - k)^2 = r^2$
- Where (h, k) is the center of the circle and r is the radius.

Ellipses

- General equation: $\frac{(x-h)^2}{a^2} + \frac{(y-k)^2}{b^2} = 1$
- Where (h, k) is the center, a is the horizontal movement left and right from the center to points on the ellipse, and b is the vertical movement up and down from the center to points on the ellipse.
- Additionally, when $a > b$ the foci are $(h \pm c, k)$ where $c^2 = a^2 - b^2$ and when $a < b$ the foci are $(h, k \pm c)$ where $c^2 = b^2 - a^2$.

Hyperbolas

- General equation: $\frac{(x-h)^2}{a^2} - \frac{(y-k)^2}{b^2} = 1$ (opens left-right)

- Where ***(h, k)*** is the center of the rectangle that contains no points of the hyperbola, ***a*** is the horizontal movement left and right from the center to the points on the rectangle, ***b*** is the vertical movement up and down from the center to points on the rectangle, and $\mathbf{y-k=\pm\frac{b}{a}(x-h)}$ are the equation of the asymptotes.
- General equation: $\frac{(y-k)^2}{a^2} - \frac{(x-h)^2}{b^2} = 1$ (opens up-down).
- Where ***(h, k)*** is the center of the rectangle that contains no points of the hyperbola, ***b*** is the horizontal movement left and right from the center to the points on the rectangle, ***a*** is the vertical movement up and down from the center to points on the rectangle, and $\mathbf{y-k=\pm\frac{a}{b}(x-h)}$ are the equation of the asymptotes.

28 Sequences & Series II

Review of Equations

- An **infinite sequence** is a function with a domain that is a set of positive integers; written as $a_1, a_2, a_3, \ldots, a_1, \ldots$ with each a_i representing a term.
- A finite sequence is a function with a domain that is a set of only n positive integers; written as $a_1, a_2, a_3, \ldots, a_{n-i}, a_n$.
- Summation: $\sum_{k=1}^{m} \mathbf{a}_k = \mathbf{a}_1 + \mathbf{a}_2 + \ldots + \mathbf{a}_{m-1} + \mathbf{a}_m$ where k is the index of the summation.
- n^{th} partial sum: $\mathbf{S}_n = \sum_{k=1}^{n} \mathbf{a}_k = \mathbf{a}_1 + \mathbf{a}_2 + \ldots + \mathbf{a}_{n-1} + \mathbf{a}_n$
- An **arithmetic sequence** or **arithmetic progression** is a sequence in which each term differs from the preceding term by a constant amount called the *common difference*; $\mathbf{a}_n = \mathbf{a}_{n-1} + \mathbf{d}$ where d is the common difference.
- A **geometric sequence** or **geometric progression** is a sequence in which each term is a **constant multiple** of the preceding term; $\mathbf{a}_n = \mathbf{r}\mathbf{a}_{n-1}$ where r is the constant multiple, is called the common ratio, and $\mathbf{r} = \frac{a_k}{a_{k-1}}$.
- $n!$ or "n factorial" $= n(n-1)(n-2)(n-3)\ldots(3)(2)(1)$; note: $\mathbf{0! = 1}$

Properties of Sums, Sequences & Series

- $\sum_{k=1}^{n} (a_k + b_k) = \sum_{k=1}^{n} a_k + \sum_{k=1}^{n} b_k$
- $\sum_{k=1}^{n} ca_k = c\sum_{k=1}^{n} a_k$ where *c* is a constant
- $\sum_{k=1}^{n} c = nc$ where *c* is a constant
- The nth term of an arithmetic sequence is $a_n = a_1 + (n-1)d$ where *d* is the common difference.
- **Arithmetic Series:** The sum of the first *n* terms of an arithmetic sequence with a_1 as the first term and *d* as the common difference is the nth partial sum: $S_n = \frac{n}{2}(a_1 + a_n)$ or $S_n = \frac{n}{2}[2a_1 + (n - 1)d]$.
- The nth term of a geometric sequence with a_1 as the first term and *r* as the common ratio is $a_n = a_1 r^{n-1}$.
- The terms between the first and the last terms of a geometric sequence are called the **geometric means**.
- **Geometric Series**: The sum of the first *n* terms of a geometric sequence with a_1 as the first term and *r* as the common ratio and $r \neq 1$ is the nth partial sum: $S_n = \frac{[a_1(1-r^n)]}{(1-r)}$.
- The sum of the terms of an infinite geometric sequence with a_1 as the first term and *r* as the common ratio, where $|r| \leq 1$ is $\frac{a_1}{1-r}$; when $|r| < 1$ because if $|r| > 1$ or $|r| = 1$ the sum does not exist.
- The rth term of the binomial expansion of $(x + y)^n$ is $\frac{n!}{[n-(r-1)]!(r-1)!}x^{n-(r-1)}y^{r-1}$.

29 Problem Solving: Summary of Techniques

Review of Equations

Odd Numbers, Even Numbers, Multiples

- First number = **x**
- Second number = **x + d**
- Third number = **x + 2d**; etc., where **d** is the common difference between any two consecutive numbers in the set.

Money, Purchases

- **V** is currency value or purchase price
- **C** is the number of coins, bills, or purchased items
- Formula: $V_1C_1 + V_2C_2 + \ldots + V_nC_n = V_{total}$

Mixture, Solutions

- **V** is the volume
- **P** is the percent solution or mixture
- Formula:
  $V_1P_1 + V_2P_2 + \ldots + V_nP_n = V_{final\ value} P_{final\ value}$

Work

- **W** is the ratio of time to complete the job together by a team of people or machines compared to the time to complete the job alone by one person or one machine.
- One, **1,** represents the whole job
- Formula: $W_1 + W_2 + \ldots + W_n = 1$

Distance

- **d** is distance
- **r** is rate or speed
- **t** is time as indicated in the rate
  Ex: miles per hour or meters per second
- Formula: $\mathbf{d = rt}$
- These relationships can be used depending on the situation described in the problem:
  - $\mathbf{d_{to} = d_{returning}}$
  - $\mathbf{d_1 + d_2 = d_{total}}$

Proportions & Variations

- ***a, b, c,*** and ***d*** are quantities specified in the problem
- $k \neq 0$
- Formulas:
  - Proportion: $\frac{a}{b} = \frac{c}{d}$; cross-multiply to get $ad = cb$
  - Direct variation: $y = kx$
  - Inverse variation: $y = \frac{k}{x}$
  - Combined variation: $y = \frac{kx}{z}$; y varies directly as x and inversely as z